GOD, WORLD, AND TIME

ANDRÉ MERCIER

GOD, WORLD, AND TIME

With a Preface by
RAIMON PANIKKAR

PETER LANG
Berne · Berlin · Frankfurt/M. · New York · Paris · Vienna

Die Deutsche Bibliothek – CIP-Einheitsaufnahme

Mercier, André:
God, world and time / André Mercier. With a pref. by Raimon Panikkar. – Berne ; Berlin ; Frankfurt/M. ; New York ; Paris ; Vienna : Lang, 1996
ISBN 3-906755-80-0

Printed in Germany

Dedicated

to all those who have been my students and more particularly my pupils during the fourty years 1939-1979 of my teaching activity,

- among others notably to the first of them, the late RES JOST†, Professor of theoretical physics at the Swiss Federal Institute of Technology in Zurich, -

as also to all those who have been, or are still, my colleagues since my retirement sixteen years ago,

especially - yet not only - at the Institute of Theoretical Physics, but also at other Departments, more particularly at the Physics as well as the Mathematics and the Astronomy Institutes of the Faculty of Sciences where I taught theoretical physics and philosophy,

as well as at the Institute of Philosophy of the Faculty of Arts and - last but not least - at the Faculty of Theology where I also taught philosophy during the last years before retirement.

University of Berne
1994/95

ACKNOWLEDGMENTS.- The Author is particularly thankful to Professor MOGENS WEGENER from Aarhus University, DK, for long discussions held on various topics of this Essay as well as for valuable suggestions, the most conspicuous of which has led to the choice of its title.

Other scholars have also contributed critical and/or constructive suggestions, for which the warmest thanks shall be addressed here.

TABLE OF CONTENTS

- o -

PREFACE
by

RAIMON PANIKKAR, Ph.D., Th.D., D.Sc.
Professor Emeritus, University of California

Only a philosopher at the acme of his career could responsibly write the *Summa* I have the honour and happiness to prologue. As he will convincingly say, echoing an almost universal tradition (everything complete is always triadic, says the Mahābhārata following the Bṛhadāraṇyaka Upanishad) "*three terms are required* to establish the springboard from which to start any original cognitive undertaking" (p. 17). *God* and the *World* would seem to be sufficient to cover it all. The fine intuition of ANDRÉ MERCIER, and here the reader is free to flare the trinitarian insight of the Christian tradition, leads him to introduce I should not say a third ontological being, but certainly the metaphysical link 'between' the two - where, of course, all these words have a particular meaning. *Time,* in fact, is the indispensable notion to establish the just-mentioned "springboard". Whereas time is a notion, space is only a concept, he will tell us, adding that time is neither a quantity nor a quality and not even a property or a value (p. 17). Time and Being are "prime notions", MERCIER says citing HEIDEGGER, and going beyond him including God in the triad.

This dense paragraph will give the reader an idea of the ambition of this relatively short book which could be read as the intellectual and also spiritual testament of this scientist turned philosopher by the intrinsic power of the two disciplines which he has mastered. And here lies the importance of this study. In our era of lethal specializations neither a 'pure' scientist, nor a mere philosopher could have written this *Summa* for our times.

These last words are also meant to convey the limits of this treatise. It is indeed a genuine philosophical reflection on how the prevalent Western or westernized scientific mentality of our generations envisages the perennial problem of the meaning of the world in which we live. This world is seen primarily as the material world as described by modern science whereby the word 'science' (*scientia, gnosis, jñāna*) has had a shift of meaning as well as the word 'world' (*mundus, cosmos, jagat*). God is no longer a given, as in other cultures. We need a jump, and MERCIER ventures a scientific leap.

Within this framework it is fascinating to observe how it functions "*in statu nascendi*" a well-trained mind in the two fields of philosophy and science. The parameters are the scientific ones in the best sense of the word. This allows MERCIER to defend metaphysics as a science *sui generis*. Metaphysics deals with the incommensurable. And while space is commensurable, time is not. Thus time is the 'link' I did refer to - word which our Author does not use.

I am tempted to call this important book a *Metaphysica ordine geometrico demonstrata*, whereby I do not have in mind just SPINOZA, but the great and by MERCIER so admired PLATO who wrote ἀεῖ ὁ θεὸς γεωμετρεῖ - although beside geometry there is also the temporal realm (cf. pp. 151, 202).

I have mentioned that the style of the book and the *forma mentis* of the Author is the modern mind shaped by the scientific ways of thinking. It is a book written for our times.

This allows him to use the logical reasoning and to practically separate the field of 'religion' (of which theology is the hermeneutics) from that of scientific reason. He acknowledges the gap between reason and faith. The hiatus is bridged by the recourse to BOHR's principle of complementarity. This paradigm allows our Author not only to make room for the Divine but also to open an ample place for art and ethics. An intriguing thought which he does not follow up, respectful as he is of the common usage of words, is the notion of 'bonification' parallel to the known methods of 'verification' (and 'falsification'). Things, rules, relationships have to be 'bonified' by experience as much as verified by observation (p. 44). I would invite the reader to 'bonify' the many rich ideas of the book. "Poetry is ... to Art what Metaphysics is to Science; yet (he thoughtfully adds) in a reversed manner" (p. 47).

The book contains such a wealth of references to Greek and post-Cartesian Western authors (the Middle Ages are overlooked) that it can also be read as a history of Western thought and as a 'recuperation' of part of its own tradition after the understandable, but no longer sustainable 'enthusiasm' for the discoveries of the modern sciences. I have qualified the word 'recuperation' because the Professor of theoretical physics goes a step ahead. He carves a new place for God in the present day cosmology. He no longer intends to prove the existence of God, because, he says, God does not "exist" but *is*, He is a Being which is not homogeneous with any existing thing, but stands at the source of everything and is the Source of all that exists. He is more ambitious and elaborates a proof of God. Yet our Author is not tempted as some have been (not with much success) to give physical or mathematical proofs for God, precisely because God is not a physical or mathematical entity. He grounds his proof on the very value of the *logos* on the one hand and on the experience "that Love, Death and Freedom 'in their togetherness' constitute the indubitable proof of God" (p. 212), although he refuses to elaborate on what exactly this togetherness is.

I mentioned a certain dichotomy in MERCIER's thought in spite of the principle of complementarity. But I would be misunderstood if this sentence were to be interpreted without that which for me is the most important character of the man and his thought: Love. Professor MERCIER - and I speak as a friend - is not only a loving person. He shows us how "Love is *the motor of all knowledge:* objective, subjective, conjective and contemplative" (p. 212).

A preface, in my opinion, has to avoid, among many other things, three temptations: to put forward the ideas of the prologuer, to criticize the Author or to extol the contents of the book. There are other places and occasions to do it. It seems to me that I have already alluded to a sufficient number of issues so as to let the reader surmise how much one can learn from it. A preface has to be a pro-logue, a fore-word, an introduction to the book, inviting the reader, that is, making the reader envious to read it. And this is what I am trying to do.

But one is tempted only by what is worthwhile, the while of a time that is "to-come" (p. 159ss.): the while the book deserves to study it.

I wonder whether I admire more the faith of the scientist or the science of the philosopher. Probably both, although desiring a sacramental marriage, which will dawn to the attentive reader.

RAIMON PANIKKAR
E-08511 Tavertet, 15th August 1995
Feast of the Connubium of Heaven and Earth
in the Human Body

Chapter I

Incentive to the Idea of God[1]

Before being born, a future infant resides in the womb of its mother as both an 'integral part of', and a 'part to depart from', her, meaning that it has 'no problem', since it has no need to think about either *the present* (food and digestion: physical needs are parts of the mother's) or *the to-come* (winning its independence: becoming a 'person' of its own). As for its past, it is part of its mother's. It will - in some future - have a past of its own, after birth. This double circumstance implies, that - once born - the child, even after coming of age, will *never* forget its mother.

One thing is thereby evident: birth amounts to the creation of time for (or should we rather say the endowment of time to) the infant. And that is the clearest existential proof that not the past, but the future: *the to-come* is the sense (in both 'senses' of the word: signification *and* direction) of time for the newly existing being: his or her - say its - time *will originate from* the future. Before birth, everything was at hand in the mother's womb, hence nothing was missing,

[1] According to Plato, an idea is an *eternally existing pattern* of which things in World yield imperfect copies. According to Descartes, ideas are the *images* of beings not as products of fancy, but *as informative of what they are the images of.* According to Locke, they constitute the objects of thought. Apparently, according to Descartes, one ought to avoid talking about the 'idea of God', since God is not among things (Hobbes), yet Descartes has allowed that use of the term (cf. *Raisons qui prouvent l'existence de Dieu ...* §2). An idea is neither a 'concept' (which is a product - not an object - of thought, as either applying *a priori* to classes of objects of thought, like in mathematics so many of them, or drawn from experience as common to classes of objects, allowing classifications and the like), nor is it a 'notion'. *A priori* concepts are *rigorously* applicable, empirical ones rest on ressemblance, not on identity. It is therefore clear that there can be neither a 'concept of God', nor a 'notion of God', but only an *idea of God*, - an exceptional one, for sure, since evidently God cannot be thought of as a thing, since the definition of a thing must attribute finitude to it; precisely this exceptionality is the reason of writing this section. Only if there were several gods could they be classified either as an anthropomorphic concept, or as an empirical one (if human beings did experience them). The Platonic definition of an idea is the most suitable one for philosophic use, and it shall be adopted here, yet in a more extensive sense than that of so-called specifically *Platonic* ideas, the list of which would be restricted.

lacking, wanting. The *very soul* of the coming infant was within the doom of the mother's judgment, i.e. no problem, no *mystery* could be said to arise for it. In other words: such mystery was elucidated 'before-hand' or, say, commonplace. Whereas from birth on, the mystery of being arises as true mystery, or as the mystery of (or about) Truth, and solving it will become[2] the very task of the person's life till death, *unless* the (quasi-) miraculous opportunity be given that person to be put so to speak back into the womb of an 'all-mother' (yet no more of its bodily mother), where to contemplate Being as Such ... and this wording speaks by itself: con-templation is the state of being within the temple (the 'womb'), i.e., within the Abode of the Great Creator to whom mankind has always given the same name: God.- That state of being confronted with the Lord of the temple (τέμενος means not only a holy enclosure, but also the pregnancy of an expectant mother) is tantamount to knowing the mystery, which then is no mystery any more, the future with both its genuine aspect as a to-come and its interpretation as a to-become being thus (dis-)solved into a present (the presence in front of God), from which it is infinitely painful to part, like it is painful for a new-born child to part from (the womb of) its mother. For *then* (a conjunction expressing the future as the to-come) the mystery of being is there again, hidden and concealed in the existence, i.e., by the coming out (Latin: *ex-sistere*) from the temple. Hence yon experience called *mystics* (but please not mystic-ism which as an -ism is mere ideology) is the one which can give, and indeed yields the purest possible (the truest if you please) *Idea of God.*

In that perspective: "Man isn't but a reed, the weakest one in the whole of Nature; yet a thinking reed" (BLAISE PAS-

[2]Notice here the necessity to use the future of the conjugation of the verb to become. By the way, the two verbs to come and to become constitute together the paradigm of the future: as seen from the past yet only seen as such, it is the to-become, whereas when seen 'from the future proper,' it is the to-come.

CAL: *Thoughts*). That thinking reed's Thought is three-fold. For it can be (or rather manifest itself as) *philosophic thought* based on 'reason', or it can be (or manifest itself as) *religious thought* based on 'faith', - philosophy and religion excluding each other 'at one and the same time', - and it can be *mystique* which is not a 'thought proper' but that experience which overcomes the complementarity of the said reciprocal exclusion, and accomplishes the upheaval of temporality as implicated each along its own way in philosophy and religion. For, the mystic experience is to comprehend the mystery of being (cf. GABRIEL MARCEL: *Le mystère de l'être*). And to awake from the mystic experience is to fall back into the temporality of existence, where there are these two kinds of νόησις: one - *logos* - seeking a *mastery over* temporality, the other - *pistis* - attempting a *rescue from* temporality. For Time is both a river and a flood of existentiality springing up from behind the to-come, as if its Creator would pump it from the Ocean of the To-Be and let it fall onto the world as Rain from some 'Heavens'. Therefore we have the tendency to *believe* that He resides in Heaven, - yet a heaven which is not the sky of the astronomers; - and the latter kind needs a Saviour if we are not to get drowned, whereas the former is the *claim* that we can do without Him by building our own ships and bridges and havens according to schemes as born from, and borne by, our experience of temporality.

These two steps - as implied by either religion or philosophy - amount to either giving up the Idea of the World for the Idea of God's sake, or substituting for the Idea of God the Idea of the World. Thus it is quite clear, that we cannot perform both steps at a time: therein does their complementarity consist. Here, 'complementarity' must be understood as that which NIELS BOHR has meant when enunciating his 'principle of complementarity' concerning microphysics, asserting that the two aspects: as a particle or as a wave, of one and the same physical entity cannot be grasped at one and the same time by any experimental means; yet here, not a physical object is implied, but thought, and its grasp is not an experimental, but a noetic one; hence thought cannot at one and the same time be philosophic thought and religious thought. A clear consequence of that 'enlightened' principle of complementarity is that there can be neither a philosophy of religion, nor a religion of philosophy or, say, that neither religion has a right to dictate anything to philosophy as was the tendency within scholasticism, nor has philosophy a right to judge upon religion, as was the tendency at the time of the so-called (thus falsely named) Enlightenment and as is again a tendency among some contemporaries.

The only justification for giving up the Idea of God and choosing to walk one's own way through the world is consequently the sufficiently satisfactory success of the enterprise; its danger, however, is not just the possibility to go wrong, but the weakness, nay the mistake, to 'believe', that we altogether can do without God: For to believe is *not* an act of Philosophic (nor of course of Scientific) Thought: Therefore a will-be philosopher who claims, not to believe in God, is simply falling into his own trap. On the other hand, the only justification for giving up the Idea of the World and choosing to let Divine Providence look after one's fate dwells in the presence of 'feeding crows' or benefactors in one's neighbourhood who help in one's existential maintenance; its danger however is not just that one could perish, but consists in the error to 'claim' that we can do without the world: For to claim is not an act of Religious Thought. Verily, both steps require each its own courage: When choosing the path of Philosophic thought: the courage of humbleness; while when choosing the path of Religious thought: the courage of certitude. Not the other way round. For, as DESCARTES has taught (and never was contested in that): the touch-stone of philosophy consists of systematic doubt; whereas that of religion resides in truthfulness.

Yet there remains on the one hand the possibility to 'think' (of, or about) God philosophically say, as the Supreme Being on Which the existence of the world depends - and this leads to metaphysics ('beyond' the science of worldly things: physics), and, on the other hand, the timeliness persists of religiously confiding in, or trusting, the world since, after all, in virtue of the Idea of God, all existing things in the world go back to Divine creation.

The Idea of God as Yon Incommensurable as Dealt with by, or Object of, Metaphysics

There is a tradition according to which metaphysics would (or should) be a kind of 'knowledge *a priori*', its source being 'pure reason'. This, we claim, is totally besides the point, actually wrong, or rather irrelevant. Metaphysics is a *science,* for two 'reasons': first of all, there is a kind of *experience* which is *meta*-physical; second, there is a kind of 'reasoning' which is adapted to the treatment of such experience. Yet - it is true - the science called metaphysics departs from the 'usual' sciences (all of the same kind as *physics*), - the reason why we call metaphysics a science *sui generis*, - owing to the fact that the ordinary sciences of the 'physical kind' are an approach (an *objective* approach, of course) to what is com-mensurable, whereas metaphysics is (also an *objective* approach), yet to what is incommensurable, i.e., the final object of metaphysics exceeds all possible 'measurements', i.e., comparisons, with standards as borrowed from the (so-called) physical world, - and this is - if not the historical, yet - the best reason to call that science *sui generis* "metaphysics".

Incentive to the Idea(s) of Time and World

Much of change can be - if not perceived directly, yet - interpreted as - i.e. reduced to - motion. But not all of change is motion. Surely: evidently, if you please, change 'happens in time'; in other words, time is a necessary substratum for (the understanding, at least the interpretation of) change.

Yet 'location in a would-be space' is not; e.g. 'getting older, ..., death' are not interpretable as motions (in space). We *must* reckon with time. Space, on the contrary, is but a - very convenient - invention of our mind to make descriptions of many temporal behaviours possible, or at least easily comprehensible: com-prehensible, i.e., after all, com-mensurable.

Time is *not* (com-)mensurable. Time is never measured, because time is *creation* (BERKELEY dix.). Therefore, if creation is the work of (a) God, then God creates both Time and World 'at the same time', i.e. together, and consequently the creation of the world goes on as time persists (from Latin: *per-sistere*), and all (genuine, say 'true') existence (from Latin: *ex-sistere*) roots in (divine) creation along with time: "to exist is to be in time". Hence World (= the totality of what exists) and Time are *'simul'*, i.e. together. Physics is fundamentally the attempt to separate them by a suitable discourse allowing to make sense out of change.

The existence of the World is the most extraordinary among all possible facts. Its prime character is, that World and Time are (or 'make') *One.* Yet World is not 'space', it is *matter* which on its own, commits *things*, i.e., Reality. *Things exist;* space does not exist, especially a will-be absolute space does not (again: BERKELEY!). Space is an abstract construct of the human (or 'divine'?) spirit. Consequently, it is not "eine Form *a priori* der menschlichen Anschauung" (said to be 'exterior', in opposition to an interior one which would be Time), while NEWTON was right in making out of time the (mathematically conceived) independent variable upon which all the (variable) properties of things constituting the (physical) world depend, whereas his proposal to consider *space* as *sensorium Dei* is correct only in so far as "*'Αεὶ ὁ θεὸς γεωμετρεῖ.*" And KANT falsely deduced from it (or changed it into) that space is (only) man's (but not God's) sensorium: *'a priori* form of human intuition', yet exterior. whereas time would be the interior form of that same intuition; which is devoid of sense, neither the interior nor the exterior having any support to lean on.

Since God is 'evidently' the sole possible Creator of the world (else the world would not 'have been' created, i.e. it were of all eternity), the question arises: Did He fetch the necessary material somewhere? The answer to that question is immediate: No, for if He had, then that material should have been 'created' already, and this is absurd. Hence God could only create the world by separating it from Himself. This requires a *will* to create which is the A and Ω of *poiesis*. Hence the world was not created *ab nihilo,* yet 'before that separation', the world was 'not yet created': Of course, this saying so is absurd, for there 'was' no time before time accompanied the world thus created. Moreover, since time is still going on, the creation of the world is also still going on. Therefore Time is not 'flowing' from a supposed Past (that is nothing but a collapse into nothingness) towards a Future (that is not yet anything); *Time is flowing from the Future towards the Past* and is therefore a succession of surprise: *ex necopinato.* Hence the 'idea of a determinism' is null, unless it is inserted into an arbitrary representation of time as reverted, that is artificially conceived of as flowing from the past towards the future, for indeed: the past is terminated, 'de-termined', *up to the present* and only till that present, yet is nothing but 'history', in other words: of no matter, except Remembering and Learning!

Incentive to the Ideas of Being and Existence

Should we ask (like Leibniz) 'Why is there something and not nothing?' or should we ask (like Heidegger) 'Why is there Being rather than Nothingness?' Probably both questions are justified; yet they do not coincide: In any case, some- (as a matter of fact: any-)thing is part of the world; and no-thing is absent from the world. Hence the first question is a query about the creation or non-creation of the world, and it admits of two positions: either 1^0 there is God's creation, or 2^0 the world is not a divine creation and is therefore existent 'of all eternity', and then 1^0: why did God 'decide' to create the world, or 2^0: how comes that the world is existent of all eternity? So far the Leibnizian problematics. As to the Heideggerian problematics, it must discriminate between *being*[3] and *existence*. Existence is 'being in time' and must be attributed to things (*res*), hence it concerns reality: thing-ness. Yet the opposite is untrue: it would not make sense to interpret being as existence devoid of time. The English language is unable to master that problematics, for it has *no word* for non-being: French *le néant*, German *das Nichts*, Latin *nihil*, Greek *τὸ ουδέν*. The English word 'nothing' is fabricated from no-thing, but a thing is endowed with existence, hence nothingness is not *le néant*. The latter is the negation of being, not of existence. Hence the Heideggerian problematics cannot be taken care of in English. Being is, in the infinitely long run, identical with God, and the question is actually: Is there God or not? but not: Does God exist or not? which is wrongly posed. And since some existent beings are the supports of all human experience, there must 1^0 be God, 2^0 even the creation of things, i.e., existent beings by God. Therefore there is God, and *this is a proof of God* (but of course not a proof of the so-called Existence of God, since existence cannot be an attribute of God, for indeed existence requires time and time is of divine origin).

This does not prevent God from - so to speak - 'delegating part of His Being'; yet Being (divine) as such has no limit(ation); it simply 'is' (yet incomprehensibly) 'not non-being'. It is the origin, the source, *fons*, eventually *auctor*, *αρχὴ καὶ Ὢν* (not just *ἤν* which already refers to existence), in short Capital Being. Evidently that Capital Being is incommensurable with any 'thing' else. Hence if there is a

---

[3]For the moment, we use the word *being* to designate what is in French called *l'être* and in German *das Sein*. However, we shall see in a later Section that also that use of a noun is liable to criticism and that one ought to speak about 'the to-be'.

possibility for Man to talk further about it, that talk - if conducted *reasonably*[4] - is the 'Science *sui generis* of the Incommensurable', and that is Metaphysics. If, alternatively, Man *confides* in it, then Man behaves as a religious 'being'. But it is clear from what has just been said, that a 'human being' either behaves as a philosopher, or as a religious thinker. Hence philosophy and religion are complementary yet on a level different from BOHR's complementarity within microphysics, since, there, the behaviour of physical objects depends on the way these objects are approached by experiment.[5]

Consequently, LEIBNIZ is right: Being (as divine) is infinite and, as God, creator - His will and His work being One: That is ποίησις. And as far as man resembles God, he (Man) is also ποιητής, i.e., creator, poet, even metaphysician, yet different from God, since Man does not create a world by separating it from himself, for if he (believes that he) does so, his creation cannot be but 'ideology', i.e., a work into which he inoculates an 'idea' which he mistakes for an infinite measure of every human undertaking, which is absurd and unreasonable, since a measure *must* be finite *and* smaller than the measured 'thing' or 'being'. Hence, any ideology must be rejected. Metaphysics *is neither* ideology, because it starts by recognizing the incommensurability of its object: Being, or God, *nor* is it of course mystics, since it does not proceed from within the temple, but from without, and does so objectively.

All created things are finite. Yet as existing 'beings', they are endowed with such and such amount of beingness which correlates them with Capital i.e. Divine Being, allowing them

---

[4] We should add: and objectively, for there is also the possibility to write *subjectively* about it, and that will be *poetry*, which also looks for the incommensurable (see later).

[5] It should be added here that the Author of the present Essay, himself a former pupil of NIELS BOHR from the years preceding 2nd World War, has had the opportunity to discuss such matters with him, in those early days as well as later in the fifties during a sabbatical leave partly spent in Copenhagen, when BOHR agreed to extend the field of complementarity to such human enterprise as Thought itself.

to partake in the infinitude of the latter: the whole world is regarded as a cosmos, i.e., endowed with Value: at least Order, and Beauty. At the same time, their finitude condemns them to be imperfect. God alone is (absolutely) perfect, and the world, including each of its existing parts or elements, is consequently tainted with anti-value like evil, unseemliness etc., and this is understandable, since by separating the world from Himself while creating it, God cannot but keep His perfection unencumbered, even though the act of creation always - by God and by other 'poets' - amounts to a break, or violation, of 'symmetry'[6]. Fortunately, God's symmetry is infinite, i.e., inexhaustible even in the act of creation, so the break does not damage Him.

We were taught at School, that 'modern science' was made possible by Greek precursors, that the ancient Greek philosophers were indeed tackeling scientific questions etc. etc. This is both the case and not the case. The very concept of time which they possibly had developed was certainly not adapted to the elaboration of 'modern' science. ARISTOTLE's physics is simply wrong. (DESCARTES' physics too, by the way, with the exception of optical refraction.) The actual development of 'modern' physics was made possible, as is known since the beginning of the 20th Century, by the acceptance of a Judæo-Christian notion of time.[7]. Metaphysics on the contrary was made possible by the Greek, notably by PLATO (more than ARISTOTLE, in fact). Yet neither the Greek, nor the Christian thinkers of, say, the Middle Ages and even later, had formulated a clear concept of the 'incommensurable', because they had not either the scientific concept of commensurability at their disposal. Thus metaphysics developed as a 'vague' science of 'Being' as such, and this did not help to make it a clear 'meta'-physics. Still,

[6]For a detailed explanation, see A. MERCIER, *Poïesis, Source et Moteur de Connaissance* (*in* Actes du Premier Colloque de *Philosophie de la Création*, Université de Paris I Panthéon-Sorbonne, 1989, Editions Poiesis).

[7]See SVANTE ARRHENIUS, *The Life of the Universe* (London & New New York 1909, in two volumes).

PLATO must be seen, for Antiquity, as the greater Authority in that respect, just as AUGUSTINE for the dawn of the Middle Ages. This reduces, indeed, the importance of both ARISTOTLE and THOMAS AQUINAS. PLATO's *Timaios* implies God as the All-Creator of the World and conceives of Time as its fundamental sign, i.e. after all its 'signi-fication' or notification. And AUGUSTINE may very well be called the founder of a philosophy of existence. CUSANUS, who stands in that very tradition, understood that space is something like an auxiliary concept helping a detailed understanding of the created world as a 'cosmos', respecting thereby the incommensurability of Being.

But let us repeat it: the world, which consists of 'existing' beings, i.e., beings 'in time', can only be understood, as a whole endowed with temporality, if it results from the breaking of the perfect symmetry of its Creator. And, finally, if mundane, meaning worldly, is the Latin version of the adjective to be attached to the noun world (*mundus*), it is remarkable that mundanity or worldliness, *and* temporality, have in modern language - and languages - both assumed the common sense of what is rather not established according to the precepts of divine wisdom!

THE QUESTION OF THE OBJECTIVITY OF METAPHYSICS

If Metaphysics is indeed a science, even though a science *sui generis*, it is, as a science, submitted to KARL POPPER's condition to be 'falsifiable'.

Of course, any such falsification, - which applies to the scientific *theories* themselves, - is not fulfilled if just happening as 'by chance' only once. It is submitted to the possibility to be falsified again and again, i.e., without the uniqueness of chance. The task of science consists indeed in the establishment of a theory or theories summing up in logical relationships the results of repeated experiences and even of experiments (EDDINGTON). If metaphysics is a science, it must yield such a basis for such theoretical construction as submissible to POPPER's postulate.

It is therefore a very serious question to ask, whether metaphysical propositions are at all submissible to such falsifications. These falsifications would require that devices, i.e., stratagems be proposed which might repeatedly show, that Being as such *in any case* reveals itself to, or comes into, human perception as finite and temporal, i.e., existing things and not as *that,* which is fundamentally incommensurable. Our contention is, that if on the one hand in many cases attempts to grasp what is suspected to belong to the incommensurability of Being indeed collapse into the mere grasp of finitude, in certain - then authentic - cases of grasping the incommensurable, no means can be found to reduce them to the grasp of the mere commensurable, yet that a number of human subjects are unable to master - shall we say are unfit for? - such grasp, just as (or similarly to the fact that) there are people unfit for sport like playing tennis, or for artistic creation like painting or even simply tasting beauty, or mastering foreign languages: at least a long and weary intercourse with such attempted grasp is necessary if it shall succeed with some sufficient approximation to resist a 'falsification' which is then a pseudo-falsification. We know that many opponents will deny such a contention, however we must then classify them, if not among the mischievous, yet as 'malefactors', i.e., as such who lack the ability, ... the power so to speak, to enter the fortress of incommensurability, just as there are many who are not even able to master the power of mathematics. Again, such contention amounts to claiming that among human beings there are such as resemble, say sort of 'angels' and such as are, say 'unfit'; and this sounds very insulting indeed; yet this is not meant to be insulting at all; it is meant as ascerting a matter of fact.

Indeed, that quite a notable proportion of people do possess the ability to grasp certain 'truths' and other people do not, does not imply that these 'truths' are untrue; it only reveals the limitation under which the human species suffers. After all, other (animal ...) species suffer from comparable limitations, e.g. in their capacity to 'think', to 'talk', to 'know', ... to 'love'. Nothing is known to prevent, that angels, in some future, be actually 'created' - either 'by God', or by (Darwinian) evolution!

For one thing is presently clear: God - if He is - cannot be made into an 'existing object of experiment (necessitating time)', since it is wrong to consider Him as existing in time, but He *is* (e.g. the Creator of Time). Hence, if Man makes God an 'object' (of a science), the arising science ('theology') is necessarily *meta*-physical, and POPPER's postulate must make room for a 'meta-postulate'. This conclusion implies that objectivity - if defined in a way such that it must allow for unconditional intersubjective communication - must yield to a new - call it - 'meta-objectivity' which transgresses the limits of finitude as such, especially of temporality in existence. Actually, defining objectivity by means of intersubjectivity is most objectionable. Much rather, objectivity cannot be but an *attitude* by which, according to etymology, a subject makes himself the master of its object by throwing it onto the plane of consideration in order to be able to notice and notify *all what is the case* with regard to that very object: Then, in case several subjects do the same with identical objects, their notifications, if serious and complete, will if notified with the help of equivalent means of communication, coincide in content, if not necessarily in words. In order to gain the greatest possible coincidence, the use of mathematical description is highly desirable. But the latter use is not confined to the objective attitude. It can also, as aesthetics and/or ethics *more geometrico,* apply to the subjective attitude of art (harmony, counterpoint, perspective ... combined with the 'violation of symmetries' as always implied in the *creation* of works of art), as well as to the conjective attitude of morals (cf. SPINOZA's *Ethics more geometrico*). In the latter respect, one should confuse neither art with aesthetics, nor morals with ethics, just as one should not confuse science with logics and/or mathematics: mathematics do not express truth, nor aesthetics beauty, nor ethics the good: much rather they support and even insure their respective successful promotion.

This may concern the Divine as well, which embraces all three moments: Truth, Beauty and the Good as well, at the level of incommensurability, for while Metaphysics treats the Divine as its object, Poetry waits upon It as its subject, and personal morals use It as the mirror in which the self as subject is seen as its own object yet conjectively, and treated accordingly. In that perspective, even metaphysics as 'theology' (this designation goes back to ARISTOTLE) is found to be analytic, even though the Divine as incommensurable cannot but fall into pieces each of which is again incommensurable with the human standards at disposal. Similarly, poetry as the art to 're-create' the Divine is found to be symbolic (instead of analytic), and self-discipline as a behaviour towards the Divine through the self, is found to be homeopathetic (but neither analytic, nor symbolic).

TIME: A NOTION, SPACE: A CONCEPT. THEIR TOGETHERNESS WITHIN THE IDEA OF PHYSICS AS A CHRONOGEOMETRY

From what precedes follows of course, that there must also be a science which investigates the mere *commensurable*. That science is evidently Physics. But it will also have to be the science of *time*, where time may have to be conceived more elaborately than just a linearly developing magnitude originating from the future and/or artificially made into a single linear variable, the direction of increase of which may be for the sake of convenience (or of tradition) formally reversed in order that this 'traditional time variable' could and thus would be said to evolve from, say, a past that is so 'old' (from geology at least millions of years, from radio-astrophysics at least billions of years), that one might for developments of the mere and 'already existing' world base the corresponding physical theories on the idea, that time is the independent, linear, one-dimensional variable t assumed to grow from $-\infty$ till $+\infty$, and attribute to it numerical values between these two extremes. Yet this attribution is in no way the result of measurements performed upon such a (linearly conceived) time. The only two sensible things to do are: (1) *Either* to consider some 'natural device', i.e. as found in nature (or in the world if you prefer) as existing and evidently endowed with sufficient continuity (i.e. by which no 'wild' changes, especially no discontinuity, meaning no disappearance of existence, can 'reasonably' be expected), so that noticeable states of that device will be found suitable, at least convenient, to be *declared* the bearers of successive numerical values: 'hours', divisible in 'minutes', 'seconds' ..., a 'zero' being agreed upon as having coincided with a memorable state, either of the device itself, or of another notable element of the world having existed: best of course - if the world has been created, - then that very creation, but who remembers creation? *Nobody whosoever,* not even one's own birth, which renders that zero-time in coincidence with Creation radically uncertain. Hence, because of the cognitive unreachableness of that 'event': something like the *birth* (rather than the death) of an unforgettable Being into existence, of a 'king of Kings' ... (2) *Or* to manufacture such a device as resembling a natural one, calling it a clock, or a watch, upon which recognizable signs as happening states will be *called* zero, ..., one, ..., two ... including intermediary numbers (e.g. sexagesimals, why not decimals? or what have you?), defining thus a calendar, preferably more detailed than just years, months, and days... .

This must make quite clear and undeniable, that time is not measurable and consequently is not to be compared with units of measure which evidently ought to be, yet cannot be, more primitive, whereas any regular measure is performed at such and such time, e.g. on the 5th of May 1994 at 3 o'clock p.m., such a measure being the comparison of a real property of a thing with a comparable (co-natural: same) property of another thing, e.g. the heat content of a material object as compared with a calorimeter. However, not all magnitudes are easy to measure. Heat for instance is not a handy quantity, for it spreads extremely quickly into the environment of its container. This is the reason why its estimation is deduced from temperature changes (and consequently why uninformed people confuse heat with temperature). Uninformed people make other, similar confusions, for instance that of mass and weight. Mass is a quantity, heat is also a quantity; temperature is a property, weight is also a property. The most handy magnitude for measurement is the length of a rod, because rods remain unchanged very long unless they are broken. The proper length of a rod is a quality! Such quality is itself measurable by means of a shorter rod; thus it becomes a quantity. A rod of wood having the same length as another rod made of iron has not the same mass as the latter. We know, since ALB. EINSTEIN realized it at the beginning of the 20th Century, that mass is always quantity of energy, - say a 'form of energy', - for instance a mass m is equivalent to the energy mc^2, where c is the (most universal) constant of nature called the velocity of light. Heat is also a form of energy (this was known already in the 19th Century owing to ROB. MAYER 1814-1878). Temperature is not a form of energy. Available energy is paid for, e.g. as heat during the cold winter; therefore money can also be said to be a 'form of energy'; but its measure is not simply a physical, it is an economical one; numerally, it changes even, slightly or not, when estimated in gold standard. As a standard for mass, one has chosen that of a proton 'at rest' (since measuring a mass in motion relative to the observer yields values depending on the relative velocity). Such values are always attributed to things as existing beings, in physics called 'bodies'. Value, quality and quantity are interrelated (see later.)

Time is neither a quantity, nor a quality; nor is it of course a property. It isn't either a value. Especially, time *is not* money. Nor is it a 'body'. We cannot say, WHAT IT IS, as long as it is not put in a relationship with other notions. MARTIN HEIDEGGER has understood - and this is one of the very important advances in philosophy which we owe him - that Being and Time are, at least for us humans, say the *prime* notions at all if we are to make sense of our - as well as of all - Existence. But one circumstance may not be ousted in this respect: It is not possible to make sense by starting with only two notions at a time, because by connecting them somehow, one will never get to anything else than defining one by the other, and consequently they will be found in some relationship of equivalence, and this is precisely what may not happen, since they evidently are not equivalent. It is necessary to establish as a fundamental principle of the theory of knowledge, that *three terms are required* to establish the springboard from which to start any original cognitive undertaking (i.e., after all, any *theory*). As an example, in order to better understand what is thereby meant, let us recall that to get to a true understanding of physical happening by the means of the 'theory' called Newtonian mechanics, its Author postulated a very first relationship which asserts that 'Force produces the acceleration of the mass (of bodies)' - more precisely: Force (acting on a body) = Mass (of that body) × Acceleration (of the said body). First after accepting that postulate is it possible to introduce in its mathematical expression such and such laws of force and to determine, by mathematical integration, the precise motions described by bodies in the course of time. This typical example illustrates the circumstance that knowledge can be promoted only by connecting three notions in a relation (which the great mathematician DAVID HILBERT, himself a theoretician of knowledge, has called an 'implicit definition'). Such an implicit definition is not a definition as we could hope to find in a dictionary. It connects three not yet explicitly understood notions and is required by what one could call a trinitarian principle. NEWTON was the first to take such a step - although he was, as far as the interpretation of the New Testament was involved, opposed to what is known in Christian theology as the 'Holy Trinity', which however is epistemologically necessary, if one is to understand the relationship between God and JESUS as *Christ*: Father, Son and Paraclete (called the Holy Ghost or Holy Spirit) are in a trinitarian, i.e., three-fold relationship which allows us to ***implicitely understand*** what God is, what Christ is, and what the Holy Spirit is; in other words: what They are in their mutual relationship. It is most interesting, that precisely that great Opponent of the religious doctrine of Trinity, ISAAC NEWTON, grounded Mechanics, - and with it the whole of Physics and *a fortiori* of Science, - on a trinitarian principle popularly, but epistemologically falsely called NEWTON's 'first law', for it is not a law (of nature). Laws within Newtonian mechanics are then laws of force, e.g. of gravitation, of friction, of elasticity etc.

Having opened that sort of parenthesis in order to call the attention of the reader upon a very fundamental circumstance about the way reasoning should be conducted according to the theory of knowledge, let us go back to our starting point at which we said, that two notions only - in our case: Being, and Time - do not suffice if one is to make sense of Metaphysics (instead of Physics, or more specifically instead of Mechanics). The main question is then: Which is, beside Being and Time, the third 'notion' we need?

There are two possible answers, depending on whether we accept to dwell within the finitude of the human situation, or whether we wish to transcend that finitude. For the former case, we have already given the provisional answer by declaring - and this is a postulate - that Being in Time is Existence, and there, the third notion used in this trinitarian relationship is that of Existence. To begin with, the Metaphysician does not know what the three terms Being, Time,

Existence 'mean', better than the Physicist knows 'to begin with' what Force, Mass and Acceleration are. The latter is helped by the mathematical explicit definition of acceleration (as a second derivative) based on 'space coordinates' and 'time variable', yet that introduces both the notion of time and the concept of space, so actually the trinitarian situation is not gotten rid of, on the contrary, it is extended to a quaternary relationship. Of course, men have the *intuition* of what such and such notions imply and cover. But their rigorous signification can only be grasped after a long and weary intercourse consisting in the repeated and prolonged application of the trinitarian relationship.

The latter case requiring the transcendance of finitude can only rest on a trinitarian relationship implying God Himself together with Being and Time. It is so to speak the metaphysical pendant to the Holy Trinity spoken of in the Bible, since the Son is within Time: a Time, where Infinitude makes room for Finitude, while Being - like an Advocate - outspeaks the Divine Word establishing what is valid. This is the way metaphysics can and must overcome the infinite difficulty of grasping the Incommensurable. We have no mathematical expression for the corresponding postulate. Mathematics has envisaged transcendance within arithmetic (thanks to the concept of the number $\aleph$) and within the theory of infinite sets, but we cannot, for the time being in any case, mathematically ground a metaphysical theory like in the case of our many physical theories, and must content ourselves with explanations as those just developed.

So, let us descend again to the level of ordinary dialectics, where we can add a few comments upon the signification of Time. This will by and by leed us to the concept of space.

Time possesses nothing and is not a thing. Yet it is not nought. For it is that without which change is radically impossible, though it is not itself just the possible. Change of what? Of the quantities and qualities owned by bodies.

How then can such a change be asserted? By comparing values of such quantities and/or qualities at successive 'times' registered on a clock. But a clock is a device: either found in Nature, or fabricated by men, in both cases again a body, or rather a complex of bodies. If God is indeed the Creator of the World, let us then call Him the Demiurge. Which is the clock used by the Demiurge? The most elaborate answer we can give is evidently: the whole world, including all beings existing 'at any time'; evidently, no human being can dispose of that device as a clock. For the Demiurge, World, Time and Change appear One. For a human being, only petty changes are noticeable; we call him an observer. The clearest result of observation is, that bodies mostly act on one another, they *interact*, as physical language says, and what observers notice thereby are ex-changes of ... of what? Mostly of energy (also momentum, electric charge...). Such exchanges are finally registered by the so-called human senses, mostly by sight, which amounts to absorption of light. Such light can be registered by observers themselves, or by the intermediary of apparatuses as e.g. photographical registration accompanied by protocolary notices establishing coincidences with numbers which establish most exactly the date (up to fractions of seconds) on the calendar-clock used as reference. Two (or more) such registrations demonstrate the change in the course of time and support, even confirm the assertion, that time is the bearer of change.

From time immemorial, *motion* has appeared as the most conspicuous change, and has been interpreted as modification of location in *'space'*. Anybody speaks about space as if there existed such a continuum as void of any 'thing'. The bulk of things is called *matter*; also that is a most ancient designation, as that which is common to all existing things; it was down to recent times thought of as pretty much condensed, even continuous even though in portions occupying parts - variable or not - of the said space, in a way such, that void space would separate portions of space filled with matter, the whole of it constituting Space as a whole, eventually unlimited though possibly finite in 'size' so that space as a whole, together with the included matter, would amount to the 'World'. And soon enough - notably with the Ancient Greek - that Space was thought of as reaching in three irreducible dimensions, - which is one more than the two dimensions needed to elaborate *geometry* - as two-dimensional - at the time when PYTHAGORAS established his 'theorem' on the understanding that precisely theorems can, and must then, be proved. Such geometry originates from *measurements* (of fields) on the surface of the earth, either in acres for agriculture (acre and agri- are nouns of common origin) or else of pieces of land reserved for architectural building. It was very 'material': a price had to be paid for an acre. Yet one could draw figures (squares, ... circles, ...) on parchment and, the latter being very thin, the concept of a plane arose in the minds of geometers (measurers of the (surface of the) earth): two-dimensional, (tacitly flat) geometry could develop as an abstract and rigorous, operational and powerful mastership of what has since been called a plane, with distances and angles as objects of the measure, conventional units being abstracted from the use of thin rods marked with intervals of length. Numbers being known already long before, including their addition, multiplication, division, mathematics arose as the most enforcing discursive method ever invented. To add a third dimension was not too difficult for the evermore skilled geometers, and thus the *concept* of (three-dimensional) space was at hand including the possibility to establish theorems for 3-dimensional 'shapes of bodies', themselves devoid of materiality.

Did ..., does 'space' exist for all that? No, if existence is being in time. However, people - whoever they are - 'believe' that space 'exists'! Of course, even measurements can be made in all three dimensions, e.g. starting from a 'horizontal' plane for the sake of architecture with the help of rods as units. But the materiality is borne exclusively by the stones and bricks (today e.g. concrete) that are put together; the 'space' they occupy has no reality of its own, it could for instance never change either form or content in time. A point as an abstract is indeed zero-dimensional, a line (straight or not) one-dimensional, a surface (flat or curved) two-dimensional, a space (Euclidean as it is called, or not) three-dimensional, and as conceived by later geometers super-space, endowed with more and more dimensions. They are never real, i.e., they never have the nature of things: beings in time. They are pure constructs of the mind: this is quite clear when we begin to talk about superspaces, whereas people at large might 'believe' that points, lines, surfaces, or 'the' space are 'real', though this is an illusion. PLATO understood that in his own way when he said that the Demiurge always 'creates geometrically'. No geometrical 'space' needs be the will-be container of (parts of) the world of things, i.e., of existing beings constituting the world. Therefore, applying geometry to investigate the Real World - which is highly commendable and most successful - does not make in any manner either a space out of World, or a world ('the world') out of (a will-be singular) space. Physics (mainly since DESCARTES, NEWTON and their followers in the 18th Century: D'ALEMBERT, LAGRANGE, HAMILTON and others) did right to use geometry as an intellectual tool with e.g. f-dimensional coordinate-space and $2f$-dimensional phase-space ($f \equiv 3n$; n being the number of particles constituting a mechanical system, e.g. equal to the AVOGADRO number of molecules in a mole). Nobody among physicists dreams of attributing any 'reality' e.g. to such a phase-space: it is not there as wrapping the world and changing with time!

Something totally different can be said, and *eo ipso* must be said about the elaboration of Relativity Theory: In the language of modern geometry, a 'super-space' - whether Euclidean (i.e. 'flat') or not ('curved') built on n dimensions, is called a variety. It has become a commonplace, since EIN-

STEIN took over MINKOWSKI's mathematical construct of what has since been a (four-dimensional) 'space-time' built as the reunion of the three space-like dimensions of Old space (falsely interpreted as 'physical', i.e., endowed with existential reality) with Time as a fourth, geometrized dimension (made into a space-like - yet imaginary - one by multiplying time t by ic, where $i = \sqrt{-1}$ and c is the - universal, constant - velocity of light), to speak about *space-time* as if that 4-dimensional variety (as pseudo-Euclidean because of the factor i) were endowed with physical reality, i.e., after all, the container of World with every thing there is, - 'is' meaning 'exists', - and to generalize it to a non-flat (non-pseudo-Euclidean) but curved variety if universal gravitation is taken into account as producing curvature: Existence of mass (bodies endowed with mass quantities 'attracting' one another owing to an assumed 'force of gravitation' according to NEWTON) is re-interpreted as 'existence of curvature': the more pronounced in space-time where (should we say 'where-when') the formerly mass thought of would be in greater concentrated quantity. In old days - for NEWTON already, - inertial mass endowed with mere quantity had to be declared identical with gravitative mass. That 'principle of equivalence' has since remained valid, with the new interpretation, however, that inertial mass manifests itself as curvature of space-time. This introduces General Relativity instead of Special (in French *restreinte*, i.e., restricted to absence of gravitation) Relativity which ignored the necessity to explain that, which NEWTON had led back to so-called gravitation forces; special relativity considers (instead of forces: say interactions) solely the electromagnetic interaction which was found to fit admirably the 4-dimensional pseudo-Euclidean frame of (flat) space-time where transformations of (the four) coordinates, i.e., (ex-)changes between various, yet relatively non-accelerated observers, are founded on LORENTZ-transformations, and no more on GALILEI-transformations.

From all what precedes, one will understand that a most fatal misunderstanding happened around the correct interpretation of Relativity Theory and why the current accepted interpretation is epistemologically wrong, viz. the opinion, that: time having apparently been as a fourth dimension glued upon the former three dimensions of a 'space' which people (wrongly) believed was the sustainer or container 'of the world', the four-dimensional variety (since MINKOWSKI and EINSTEIN) called 'space-time' (in German originally *die Raum-Zeit* as a feminine, in French *l'espace-temps* as a masculine, in English a neuter), became considered as the quintessence - in German *der Inbegriff* - of the existing world as a generalization of (3-dimensional) *Space*, but not of (1-dimensional) Time.

Epistemologically (i.e. from the point of view of the theory of *knowledge*), that has been a fatal error, from which we still suffer at the end of the 20th Century. Let us now see why?

A linguistic distinction of epistemological nature should be made between a *notion* and a *concept*. In a, say popular understanding, these two words cover significations that are seemingly nearly infinitely near each other. Actually, they apply to quite different senses: A *notion* implies the moment of discovery of that which is concealed but ought to yield a comprehension, an intelligence: it is intimately related to intelligibility. A *concept* implies the moment of the invention of that which is productive and ought to yield efficiency, and is therefore intimately related to possibility (as the nature of powerfulness).

Time is not a concept, but a notion. Especially, it is the support of the 'existence' of World i.e., of all things.

Space is not a notion, but a concept, and does for that not possess existence.

Therefore geometry (and more generally mathematics) is a power. There is on the other hand no chrono-metry as a counterpart of geometry. Especially Time is *never* measured, and the word (to be found in dictionaries): chronometry is a linguistic misuse. There are devices called chronometers: they do not measure time, they indicate time as related to those observers who owe them; this is quite different. They ought to be called *chronodigms* (from Greek *δείκνυμι*: to show, to indicate).

By allying, by combining space with time in Relativity theory, time has not - as wrongly interpreted in our 20th Century - been made into a fourth dimension of a Concept due to serve the interpretation of the World, but (old-fashioned, 3-dimensional) space has been made into three supplementary dimensions of a Notion due to enforce the intelligibility of the World, a World, whose notoriously temporal existence is more intricate than what an assumed uniquely one-dimensional happening comprises.

Indeed, we can say that we understand today that Time *behaves* as if endowed with fourdimensionality. And this is what formerly Special, later General, Relativity implies. Therefore such theory is a space-like Theory of Time, not a space-like theory of space in which time is destroyed as such and made into (a fourth dimension of) space. Of course, it can, or may, for that, be the frame given to a theory of the world as *space-time:* When we recall the origin of that locution, first published in 1908 by MINKOWSKI in German, then translated into French and English, the other (existing) locution *(der) Zeitraum* (grammatically a masculine word) could hardly be used, for it means univoquely an interval ('*raum*') of time ('*Zeit*'). Therefore (most probably) MINKOWSKI forged the (grammatical) feminine 'die Raum-Zeit' (with a hyphen), however not giving it the meaning of a generalization of 'Zeit', but of 'Raum'. Actually, it was a good combination with the right gender (feminine like *die Zeit*) and ought to have been understood as a generalization of

Zeit: Time as *the* independent (now 4-dimensional) variable upon which everything else depends, just as in NEWTON's mechanics all properties of matter (quantities and/or qualities of bodies) depend, with *the* difference however, that the new theory does no more look for properties of *bodies* as solutions of differential equations 'of motion' to be solved, it looks for properties of *fields* (as four-dimensional, tensorial magnitudes), again as solutions of 'field equations' also to be solved.

The French and the English locutions as used then were *l'espace-temps,* a grammatical masculine, resp. *space-time,* grammatically neuter, putting both terms of the compound in the same order as in German. This may be correct for the English, whereas the French ought, from current linguistics, have them in the opposite order and speak of a *temps-espace* (just as a tiger-cat is called in French *un chat-tigre*), though phonetically this does not sound as nice. They could very well both have been understood as designating a new interpretation of Time. But they were (of course) not either, keeping the original interpretation as advocated by MINKOWSKI and EINSTEIN. Linguistic usages are hard to overthrow; whereas obsolete interpretations should vanish. Since we cannot (see above) call physics chrono-metry, we might perhaps call it chrono-geo-metry (and why not as a noun free from hyphens: chronogeometry), based on the historical origin rather than substituting chaos (indefinite extension), topos (location) or choros (limited extension) for *gè* (bodily rather than earthly - extension).

Still, in order to emphasize the correct interpretation as explained, we shall reverse the components of the compound and speak from now on about *Time-Space* (Time made intelligible by spatialization) instead of traditional space-time (as wrongly space with incorporated time).

Chapter II. The Metaphysics of Time, World and other Items

THE FOUR (POSSIBLE) APPREHENSIONS OF TIME

Nobody will deny that 1^0 there are beings, 2^0 that what there is eventually changes and consequently that there is an 'after' which differs from what was 'before' in a way such that each after becomes in its turn a before, each couple of a before and an after constituting an element of that which must be said to support the succession of such couples - i.e., of change - calling this support: Time.

Being in time shall be called existence, from the verb *ex-sistere* which indicates the change, viz., the out-(*ex*) coming from some - unknown, or known - state into sight (*sistere*), and this state admits of two characterizations (from *cha-raxo, ...ax, avi, are*: to scratch, to write, to stamp), - let us call them c_{fu} and c_{hi}, - as well as of two attributions, let us call them a_{pr} and a_{wo}, - one characterization being that the said change is a to-come as from a future, (from Latin *futurum:* the to-come) that vanishes into the past (from *patior,... passus sum:* suffered, endured), the other characterization being that the said change is a historical becoming (historical: from *historie*: known and/or told circumstance - *be* (as a prefix): about, adding notion of all over, or of excess, and *come*), i.e., the past is enlarged into what will, or shall, be; - and one attribution, a_{pr} being that the change operated concerns one and only one being at a time , i.e., each being having its own or 'proper time' of existence, the other attribution, a_{wo}, being that all beings share the same time which may be then interpreted as common time; yet since all beings in time constitute the world, we can also call it the time of the world or 'world time'.

According to these possibilities (two characterizations, two attributions each), we get *four notions of time:*

(1) the to-come proper $t_{fu_{pr}}$, i.e., the future of one single existing being
(2) the common to-come $t_{fu_{wo}}$, i.e., the future of the world
(3) the be-come proper $t_{hi_{pr}}$, i.e., the history of one single existing being
(4) the common be-come $t_{hi_{wo}}$, i.e., the history of the world.

Shall we agree that the adverb 'always' applies to a time t having lasted since its own origin, whereas the averbial locution 'for ever' applies to a time t lasting until infinity (or at least to a 'final time' then called the eschaton).

It may be that one (or several) existing being(s) has (or have) existed during an infinite time. The invention of a clock is usually assumed to allow to attribute numbers to time t, be it t_{fu} or t_{hi}, although in praxis mostly t_{hi} is attributed such numbers.

The fixation of time zero is arbitrary. There are possibilities to have the notification of a universal determination of a zero: if either there has been a creation of the world for the common time t_{wo}, or for each existing being if the same has been (or will be) created, i.e. for each t_{pr} individually, or also if among the existing beings there has been (or is) an exceptional one upon which all thinkers interested in time (and that includes in praxis all men 'at all time') can agree. In fact. a quasi agreement has taken place, first among the inhabitants of the West on the Earth, then also up to now as adopted by most of its inhabitants, namely the 'birth' of JESUS CHRIST, for which by the way not zero, but *one* has been chosen (and moreover an error of a few years has been made by those who in the Middle Ages worked on that problem). It is true however that adepts of Islam prefer to choose the *hegira*, and that under the French Revolution, an attempt was made to choose the date of that Revolution instead.

Yet, if all individual existing beings have each their own proper time, any unique choice remains aleatory, unless a procedure can be found to *compare* among themselves all proper times of existing beings: This would be *the most acute* condition for the establishment of a rigorous objectivity. Whether it is possible to fulfil that condition is not known today, for we do not possess a rigorous and all-embracing 'theory of time(s)' which should be verified without the least doubt. (It is even doubtful, whether such a verification can at all take place!) Hence, absolute objectivity is unreachable, at least 'for the time being', and probably at all. The reason for that uncertainty is, that one does not possess an indisputable and final procedure to establish what a human intuition would wish to call 'simultaneity'. If all proper times were identical, they would also coincide with t_{wo} (be it, by the way, as t_{fu} or as t_{hi}). In that case, there should be no problem about simultaneity. From a scientific point of view, however, such a universal identity must be submitted to the condition to be demonstrated, and such a demonstration requires either that there is a theory of simultaneity which can be experimentally tested, or that there are alternative such testable theories among which experiment can isolate the right one. But the only circumstances (but one! see further down) under which such comparisons have suggested themselves were for centuries, even millennia reduced to cases where the 'change' accompanying each being to make it an existing being consists of a 'motion' of its whole or of its parts, i.e., alleged changes of place and situation 'in a *space*', including deformation like becoming larger or smaller etc. as kinds of motion. However, this would make sense if and only if space itself did 'exist', yet not as an existing being to which then its own proper time could be attributed, for then it would call for a comparison of the time proper to it with the time proper to another very particular being as envisaged to make the comparison. And thus the problem 'would not be solved' but led back anew to a same problem, this:

recurrently to ever new 'same problems', and the problem of simultaneity would have no 'spatial' solution. However, a main mistake is made when attributing existence to space; for space - whatever 'particular space' is geometrically, i.e, within mathematics but not physics, thought of - is never an existing being, it is a concept as invented by the mind of human beings - and possibly by God. Therefore space *is not* a form *a priori* of our, i.e., human, sensible intuition, and it is further especially neither a thing like other existing things, nor even a notion as grasped by intuition, but a pure (and mere) concept; consequently, motion can only be defined as relative motion by arbitrarily declaring, to begin with, one existing being as 'being at rest', and by an agreement upon, then, how 'distances' and 'angles' towards a moving being shall be determined: 'measured'. Under such conditions, the being (said) at rest and the other beings are in physics (more specifically mechanics) called 'bodies'. In short: bodies must have a 'means of reciprocal communication' if they are to estimate their relative motions, and such a means must be universal. But there is no absolute motion as spotted by anyone with regard to *a* universal space within which everything would 'exist'. For it would have to 'have been' before anything was created, i.e., before there was any time (and in particular any change), and this is a contradiction in terms: There cannot be time before time, especially for the purpose to ascertain that there 'was' space. In a way, space is nothing-ness, since it is thought of as the vacuum, i.e., the absence of anything.

Now then, apart from motion ('in space'), what could be the nature of some 'existant' as fit to show that there is a change? The answer is: There is, and there is only, *light*, showing for instance a change in colour, but showing even also 'motion' as if it were a change of position in space. Any other means, e.g. sound, taste, odour, contacts ... need a medium to establish the relationship with the thing undergoing change, intermediary between the changing thing and the observer of change, whereas light 'travels' through vacuum, and is the only 'thing' known to do that, not needing a train or a jet or a boat.

'Light', i.e., electromagnetic waves composing a whole spectrum, is of course also an existing kind of being. But that being has exceptional, even radically exceptional, properties: it can by human artificial means be created and annihilated, and is 'seen' (either by eyes of living creatures or by artificially fabricated devices called photographic plates or films). Light seems to be a most singular thing among all other existing beings: its electromagnetic theory, which is the best verified of all theories so far proposed by physics (in fact still better than Newtonian mechanics, which had to yield to it at the beginning of this Century), asserts that when interpreted as travelling from bodies to bodies, it consists of electromagnetic waves (which can also be interpreted as particles called photons) animated by a *velocity* c (practically equal to 300 000 km/sec), the square of which equals the ratio μ_0/ϵ_0 of the two natural constants involved in that electromagnetic theory, called the permeability, resp. the permitivity (also dielectric constant) of vacuum. (We leave refraction of light through media undiscussed.) That velocity c is therefore itself a 'universal', natural constant - universal, i.e., which does not depend on any change (of anything) and is therefore found the same by measurements performed by any observer, whatever its proper time be. That velocity is so big that, even though one can in the abstract think of a still greater velocity, no existing being, supposedly moving, could ever be found to travel with a greater velocity. In other words, light is able to catch and overtake any ('flying') existing being. Hence light is unique among existing beings. It is most fortunate, that nothing can be found moving (relatively to any other thing) with a velocity greater than such a maximum, because otherwise there might be infinitely fast relative motions which would make one simply crazy.

NEWTON had tacitly assumed that time is unique as a universal t_{wo}, all t_{pr} proper to individual bodies being equal to t_{wo}, as far as the same zero is agreed upon for all and he only conceived of a historical time $t_{wo_{hi}}$. This had the practical advantage, that the problem of simultaneity was automatically solved. (Probably NEWTON did not think at all about the said problematics.) Yet he moreover made an assumption, which tacitly ran against the then accepted view of a beginning of the world (by divine creation), namely the assumption, that

$$-\infty < t_{Newton_{hi}} < +\infty,$$

in other words, that the world has always existed and will exist for ever!

The possibility that the world had been created was thereby from NEWTON's time on extracted from scientific consideration - the corresponding narratives as contained in the holy Books of any religion about the Creation of the world, e.g. in Greek called *Genesis* in the Old Testament, had to be discarded. This opened the way toward the doctrine of determinism, although determinism is, contrarily to LAPLACE's view, not a necessary consequence of (Newtonian) mechanics, since it is conceivable that the so-called world remained empty of any being susceptible to enter existence i.e., temporality between $-\infty$ and a time t_{cr} 'of creation' not infinitely, though very distant, in the past from e.g. the date of publication of NEWTON's *Principia* 1687.

The said (tacit) hypothesis has a great disadvantage, viz. that according to the second principle of thermodynamics (published 137 years later by SADI CARNOT), the 'world' ought to have already reached a complete state of chaos, since any 'now' (even one having happened long, long ago) is infinitely distant from a still more infinitely remote, already prevailing existence as of an (historical) time devoid of any beginning. This, it seems - unless there has been Creation - can only be compensated if, again and again, 'life' is infused into the world. Now if on the one side there should have been, in the Newtonian picture, no such unique event as 'creation', such infusion seems in the same picture not impossible, if there all the same is a special kind of creation (not to mistake for evolution according to DARWIN) - by whom then? - (there is only God the demiurge as a possible answer!) in the form of a quasi-continuous phenomenon. In such a case, then, the use of the picture of a 'to-come' is much better adapted than the picture of a historical time, for, indeed, what 'is to come' roots in the very potential of creativeness: especially at each moment of t_{wo} (in case we keep to the Newtonian idea of at least the overwhelming importance of t_{wo} over many t_{pr}'s that could bear each their autonomy).

In this new picture, the zero of time t_{wo} ought best to be chosen anew at each instant as coinciding with that very instant, leaving it to a t_{hi} to follow up all these 'successive' new beginnings of the to-come... or, if one wishes at all costs to keep to one single zero like the birth of JESUS (minus one year!), to give negative (increasing) values to the to-come, increasing positive values designating then the ever more remote past with regard to that chosen zero.

The great advantage of dropping NEWTON's tacit hypothesis of 'no creation' (even by steady state of infusion), resides then in the possibility to rescue the second principle of thermodynamics in view of the fact that the world is evidently not (yet) a chaos (assuming its 'beginning' to have taken place at a not too remote past), and the advantage of using then the to-come (rather than t_{hi} which is a 'to-become') is not only to give a highly more vivid picture of the possibility of a continuous creation if there is such, but also (a) to meet the danger of the tendency towards a future chaos since continuous creation amounts to infusion of (ever new) life (defined as the possibility to resist the tendency towards death as involved in the 2nd principle of thermodynamics), and (b) to allow for a 'decision' of the demiurge to stop eventually creation, to cut so to speak that invigoration at a future time then called the *eschaton,* possibly - but not necessarily - making it (as some interpretation of religious texts will have it) a catastrophy amounting to 'the end of the world' and, since Time (as t_{wo}) and World belong together, from the beginning till the end of the to-come, an end, which is then typically not in the past, but in the future. However, since we do not know by science (1) whether indeed it will come, (2) when it will come if it comes, we must leave unanswered physical questions about its happening, just as we also must leave unanswered questions as to the speed with which a final state of chaos could be reached if no such invigoration did happen, for we have no theory about how quick the disorder implied by the second principle progresses, nor do we have experimental (or at least observational) token about it.

By the way: to shift from t_{hi} to t_{fu} has no mathematical influence upon the possibility to apply any of the current physical theories, especially Newtonian mechanics, other than to replace infinitesimal increases of time dt_{hi} by decreases $-dt_{fu}$, so that the integrability of differential equations still prevails. Especially, the mathematical rôle assumed by time, in all such theories, as the independent variable upon which variable physical magnitudes will be found to depend, obtains along the solution (by integration) of the fundamental equations of each theory, where care must be given to the attribution of right values to the magnitudes at 'limit conditions', e.g. at 'time zero', which remains the sole critical choice to make in a way such, that 'historicity' be made an exact business.

So far, as to a time that would be the common time of the world and of the existing beings (including beings coming into existence along the to-come), the problem of simultaneity *not* being posed, since it is conjured away in the hypothesis of the identity of t_{wo} with the many t_{pr}'s.

But a question can - and must - be posed: Are there positive indications of a continuous (or continued) creation along the to-come? The answer is: 'in a way' yes: they are all the events which have the 'right' to be designated as births. By this, we mean the occurences of the coming into existence of beings which cannot be completely explained by a deterministic procedure.

The usual idea of birth is that of the coming into existence of a being procreated either by asexual generation, or by sexual reproduction. Most of the known cases of procreation are of sexual nature, and these require the meeting of male and female, an event which is not necessarily submitted to strict determinism, and which - if it takes place - does not necessarily end in the generation of a descendant.

These two modalities (asexual and sexual generation) are ascertained by science. In former days, science also envisaged the possibility called spontaneous generation out of inanimate matter. That third possibility however seems excluded from what happens on 'our' Earth *in the present conditions;* however, it is not possible to exclude it for all times past (and probably future times too), for else, how could life have at all appeared on Earth? (These considerations go back mainly to PASTEUR.)

Contrarily to physics proper, biology as involved here does not apply mathematical procedures to elaborate one or the other 'theory' which could be verified or falsified by observation and/or experiment. It is therefore not considered an 'exact' science, even though its sayings may be 'universally' accepted, i.e., by all presently living competent specialists.

Most reproductions in Nature are of the sexual kind. The asexual kind happens by bipartition, sporulation, budding, and similar procedures of a single, banal cell. The sexual kind requires the union (called fecundation) of two specialized cells called gametes (from the Greek *γαμέω*: to marry) sexually different, though sometimes identical (isogamy), sometimes possessing both the two sexes at a time. Not every gamete meets in due course its other gamete to continue the existence of the species before it dies, it is a question of chance, for which the probability may be very big, and not even that probability is warranted by any determinism, because the bearers (e.g. the human bearers) of the gametes may decide to copulate or not to copulate even though sexual love may waken their reciprocal attraction; and then if they do, it is still a question of chance whether the gametes actually meet during the sexual act. If it happens that within a given species no such meeting of pairs of gametes takes place before all individual bearers of the gametes (at least of one sex) die, then the species disappears (this is then determinism *à rebours:* the wrong way), - which can be due to a cataclysm acting as a sort of external determinism, i.e., not bound to the species itself. This has presumably been the case for extinct species like dinosaurs etc. There was once a school within the geological sciences advocating a theory according to which repeated destructions (by cataclysm) of all life on Earth were followed by new creations (by spontaneous re-generation, if not by the demiurge - which is considered an un-scientific explanation). The Deluge as told in the Bible is not exactly the same, since NOAH is said there to have 'saved' a pair of each species (except Man, represented by several) to prevent final disappearance of life.

In any case, the period of time required for the appearance, from a supposed spontaneous generation of at least one most primitive living being (cell etc.) and, as a consequence of evolution, of a multitude of evermore superior species represented by important populations on the Earth, is by geology and more specifically by paleontology found considerable, estimated as it is in millions of years - even billions (for the pre-cambrian period from which some organic vestiges of life seem to have existed as algas, sponges, mollusks, worms), inspite of the disappearance of the presumed continent called 'Gondwana' (where there is now the Pacific Ocean), resembling a cataclysm for which one has no final explanation, the current one being that the Earth, in those days still flabby, took in consequence of its steady rotation the form of a pear, the small portion of which, composed of light, superficial material, being then torn away by centrifugal effect and becoming the present Moon, leaving a huge flat wound in the shape of the Pacific Ocean (this theory goes back to mathematical, geodetic calculations made by JAMES JEANS and was further advocated by HAROLD JEFFREYS and ARTHUR EDDINGTON).

As is clear, the tendency is to avoid any hypothesis that would sound 'unscientific', though it is very difficult to make precise, what would be unscientific. Especially, the hypothesis of a 'spontaneous' generation has still the taste of a concealed, say theological explanation. In order to avoid it, one has taken recourse to the notion of chance, i.e., the way things fall out in the absence of design or discoverable cause (as the *Concise Oxford Dictionary* has it), for which finally a mathematical frame has been elaborated in the Theory of probabilities, able as this theory - founded by BLAISE PASCAL, 1654, worked out by FERMAT, 1654, HUYGENS, 1656, JACQUES BERNOULLI, posthumus 1713 - has been to say something certain about that which is only probable. This was the course followed at the end of our own century by PRIGOGINE in showing that such probability concerning the ordered coming together of organic elements from a disorder does not vanish. In other words, the idea of spontaneous generation: unfit, unpresentable in the 'good society' of seemly scientific theories as it was when looked upon in its original clothing, has been made presentable to scientific consideration, transformed as it is into one as is supported by a respectable mathematical concept.

This does not mean however, that myth has been transposed into science, but that progress has been achieved in the formal understanding of processes going on 'in time and in this world'.

Now, this time which is our concern here is the time during which lots and lots of beings exist, on the Earth 'tis true, and far beyond the few laws of interaction known to the most modern physics, even beyond chemistry; yet any question as to a reason for the chance of such happening is left totally unanswered, because scientifically spoken, one does no more need such, as soon as the answer has been substituted by a scientific, if possible mathematically expressed theory. It does not claim more than e.g. gravitation did within, say, the Newtonian theory. Yet the latter has been after three centuries replaced by General Relativity theory. So there is no certainty that the afore-said will not have to yield to another explanation.

The Meaning of World

Just as Time (and other nouns, like God) have been written so far with a capital, dropping the definite article, we could write about World as written with a capital but without the definite article: there is the one world, called World, but not one world as e.g. chosen among possible worlds like those thought of by Leibniz.

It could be defined as the totality of what 'exists,' though this could have one disadvantage, for: would it have to include Time? or is Time 'something else' than World?

In any case, World is not the totality of what (there) 'is,' for, assuming there is God, then World would include God. It would need a 'theory' according to which World identifies itself with God ... or God with World. Such an identification sometimes receives the name of pantheism. In modern times, ALB. EINSTEIN has more or less shared the view that the 'world' which is assumed to be (completely) described and explained by the most General theory of Relativity as supposed elaborated in its complete and final form (especially including the realisation of the Einsteinian dream of a 'unitary theory' of all fields, i.e., a theory within which all working interactions would be formally integrated), identifies itself with God thus conceived of as 'all-god', i.e. from the Greek: *pantheos*. (It would then include Time, since in the Minkowskian interpretation, time is incorporated into the four- or more -dimensional variety called space-time.)

We should realize at this point that such a view is the very last, say ultimate possible stage of what is known as *animism*, for, according to animism, any-*thing* like a tree, a hill ... a celestial body ... is animated by a 'soul' of godly nature, and if World is that which includes the totality of such 'things' (as existing), then World has its own soul, and that would be *the* god, no other smaller gods being envisaged. Yet that god would (have to) 'exist', and this is in contradiction with what has been said so far, namely that strictly conceived existence is not a possible attribute of God.

SPINOZA had already a similar conception as subsumed in the formula *Deus sive natura*: 'God or, what is the same: Nature.' There, Nature stands for World, as absolutely and insuperably autonomous, the totality of all singular existing and material bodies, which however need not to have been, hence were not, created beforehand.

For EINSTEIN, more specifically, every thing, including every human being ('human' implies 'existing') must then be declared identical with its world-line, or more precisely with the bundle of world-lines of all the constituent parts of its body.

And these make together a sort of universal tissue from which 'determinism' is eliminated by the circumstance that, owing to the Minkowskian interpretation, time collapses as an autonomous notion into the geometrical many-dimensionality of the concept of space-time, - this being, however, in our opinion, the *wrong interpretation,* since not time should be made into a space-like dimension, but space as a means of formal interpretation has to be incorporated into the many-dimensionality of a Time that is and remains *the* independent (yet super-complex) variable upon which all the constituting fields of the world depend.

World is therefore *not* a mere complex whole imbued with spatiality from which temporality can be singularized by isolating one individual, so-called observer. Much rather, World is actually a huge set of beings embedded in a complex temporality in which each being is endowed with a temporal nature consisting in *its capacity to meet the to-come.* Nothing prevents one from assuming, that these beings are the outcomes of Divine Creation, further that their temporality is also of Divine origin, that they are all subjected to the necessity of encountering the To-come, especially to a decay which physics subsumes under the second principle of thermodynamics, and even that the Source of that To-come is no other than God Himself, yet as *deus absconditus* for sure, who absconds 'behind' the future i.e. behind the to-come, in any case not in the past. To object to this assumption that it be gratuitous would be unsatisfactory, because no other justification of Time, and especially of Change, is known. And then, assuming such creation, World and Time are being created *together*, whether suddenly and once only, or continuously since an undefined or unknown beginning.

Moreover, contrarily to a rather popular belief shared by many a scientist (we remember, that philosophy, hence all the more science, should refrain from any 'belief'), a final, complete and exact, i.e., mathematical description of World is far from being at our disposal; at the end of the 20th

Century, only those 'phenomena' included in the theories elaborated for electromagnetism and gravitation (together: in General relativity) are very satisfactorily mathematically described, those depending on the so-called week and strong interactions being (only) pretty well incorporated in mathematically organized (e.g. ten-dimensional) theories, chemical phenomena also, ... but a huge field is still open for such theoretical foundation, quite specially all what is subsumed under the concept of *life*, excepting *death* insofar as it is one manifest verification of the second principle of thermodynamics. Yet all scientific progress made so far seems to confirm the Platonic adage according to which - if there is Him and if He creates World and Time - *God* (then to be called *the Demiurge*) *creates according to mathematically describable patterns of action.*

Now, since we have just put the fundamental hypothesis of a Divine creation in the *present*, and not in the past, tense, the following comment is desirable. According to the picture put forward, Creation is not just an act both momentous (for sure) and momentary (i.e. of no duration: in our days often pictured as a big bang); on the contrary, it is - if surely momentous - yet not just possibly, but even in a way clearly prolonged action, which surely may have had its exact beginning in coincidence with the beginning of Time or To-come, however which continues along that time called now the to-come, so that those beings of the world which are 'still existing' experience the said to-come as a sort of shower or rainfall of temporality which has not only the nature of e.g. a four or more dimensional, complex geometrical variety, but is accompanied by all the signals of that to-come like a stream of surprise or surprises. Especially, the concept of a determinism is not simply totally insufficient, it is unfit to account and/or reckon for the to-come; and indeed it has been eradicated first from micro-physics (HEISENBERG, BOHR) and consequently, strictly speaking, from theoretical physics proper altogether. In other words, according to present-day physics, the past does not determine the future; much rather the present which, by cutting the past from the future, seemed to classical thought to be thoroughly, even radically prepared by the past and hence pre-dictable from a thorough knowledge of the latter, is actually the immediate receptacle of the to-come which very well could deliver into it an event bearing a totally unexpected content, which needs not be interpreted as a wonder, for it happens again and again. Much rather therefore we should say that the future, i.e., the to-come implicates - or shall we say fills up by and by (French: *emplit*) the past with nothing but a 'history' which is 'finished', that is, terminated and if you wish 'determined', and is consequently unchangeable anymore, even though eventually recorded in files and books, as well as in the memories either of 'people' (but these are mortal!) or of hard disks and diskettes prepared on computers and the like and can be recalled e.g. to later historians who may wish to analyse them and sometimes write further books 'about the past'. But again also these will have to die; this is part of their proper to-come, even though it very often looks like a consequence of past events (illness, accidents ...); but diseases as well as accidents fall upon people as the marks of the to-come inspite of looking like the outcome of the past and consequently as predictable. Such predictability has stirred the minds of thinkers since time immemorial, filling them not just with an awe, but also with the yearning and hope to be able to *make predictions.* Even mythologies have eventually included the possibility of omens as interpreted by augurs, until they were replaced by scientific argumentation as allowing for the exact prediction of a variety of phenomena, yet practically restricted to mechanical explanations of motions, be it e.g. in ballistics, or about the celestial bodies. But what such very successful explanations reveal is that the to-come itself is extremely well and regularly organized, as if Divine creation would indeed imply systematic regulation, in other words a to-come as dominated by mathematical purposefulness finding its cosmic (i.e. both universal and beautiful) expression in the Platonic adage already quoted further above about God the Great Geometer-Demiurge, yielding a support to the doctrine called Platonism. Yet nothing worldly can prevent a divine decision to break at any future time the regularity of the said geometric course, even if men cannot interpret such breaks in the geometric purposefulness otherwise than by speaking about 'miracles'. For, after all, the *greatest* miracle consists in *(divine) Creation* itself, this assertion being the only possible and meaningful answer to the Leibnizian question already mentioned: why something rather than nothing?

And, coming back to the subject of death, we conclude at this stage that death is, after all, perhaps the only predetermined (pre: from the future, *not* from the past) character that can be attributed to human beings with a certainty which is confirmed by the surest experience they have of both World and themselves, i.e. of their fellow-men, yet at the same time themselves bereft of the possibility to 'mathematically predict' its coming in any single case, unless by killing someone on the spot, which puts a problem to ethics, but not to mathematics.

It should be clear by now that World as the most extensive and comprehensive existent 'being' should not be confused with a model as produced by a physical theory (or eventually by an ideology) about an order which seems to rule the formal arrangement and behaviour of bodies *in* the world. Orderliness, especially conceived of as beautiful, is the sign of a *cosmos*, and the corresponding part of science will quite naturally be called *cosmology.* However, cosmology is not the all-embracing knowledge about World (still less is it the full comprehension of Time). SPINOZA's *Ethics* for instance - which is not to be considered an ideology, for sure - is not part of cosmology, even though it be presented to the reader *more geometrico*, especially since ethics are to *morals* (i.e. to the choice of rules of *good* behaviour which should rule reciprocal behaviour among human beings) what mathematics are to *science* (i.e. to the establishment of *true* relationships which seem to rule the natural interactions between constituents of World). In drawing up this comparison, both the goodness of rules and the truth of relationships are to be ensured, i.e., 'verified' by experience (the word 'bonified' does not exist in either the Latin or the English language), and we know that experience can in both cases lead at most to some approximation of the ideal looked for in such enterprises. Hence we must in either case content ourselves with what we reach in our attempts. Sometimes, conclusions arrived at break down, not only in the domain of morals, but also in that of science (e.g. in the case of the Aristotelian physics). Perhaps the prevailing opinion in our days and in the West is that Science has reached a much higher degree of verification than Morals (even in their various codes of jurisprudence, which by the way vary quite noticeably from State to State), and then the same opinion may develop into a 'belief' that Science 'should' play a leading rôle and/or even overtake the task of establishing codes of good behaviour. Yet 'should' is a postulate of moral nature, not of scientific nor simply of logical nature. And for a comprehensive analogy to be, that comparison ought to envisage a moral approach encompassing not just mankind, but all components of World. (This is what a totally universal and comprehensive 'ecology' might do; yet then its designation as ecology is ill-chosen; even eco-nomy is not very satisfactory, though better for sure; but then the word - current in modern language - 'astronomy' ought to be replaced by astrology, which is made impossible by the contemporary usage of that word!

Similar considerations could be developed about art as the creation of beautiful arrangements ruling the apprehension of World (including mankind itself) by the senses; yet in our days the condition of beauty is being dropped by many a will-be artist, possibly as a consequence of a similar contention according to which a (pseudo-)scientific procedure of imitation and invention would suffice to the production of artistic stuff, instead of realizing, that the human attempt at reproducing the divine act by a human act of creation has to also go the 'way of geometry' followed by the Demiurge as was announced by Plato, in the understanding that the analogy between mathematics and ethics must be extended to aesthetics as well: *more geometrico.*

Under such circumstances, the question arises, to find out whether one can be satisfied with a mere physics of World (and Time), even leaving aside the parallel approaches of ethical and aesthetical nature, or *whether* - owing to a genuine incommensurability as rooted in World (and Time) - *there is an authentic metaphysics of World (and Time) worth tackling*, extending that question to the problem of the triple nature of that incommensurability as approached along Science, Morals, and Art.

Attempt at drafting an Abridged Vocabulary of Terms as used to establish Relationships between God, World, Time and Further Items

1. *Argument.* - Here is first a *thesis* as confirmed by the History of language and literature: "A language approaches its perfection and stability through the usage *philosophers* **and** *poets* make of it". This was already the case in Ancient Greece.

This became especially the case when (Greek and) Latin were by and by abandoned (including the use of Latin at Church from the Reformation on and after), and philosophy - especially metaphysics - and even religious thought - were written in French (Montaigne 1533-1592, Descartes, Pascal, Bossuet, Leibniz, ...), in English (Locke

1632-1704, BERKELEY 1685-1753, ... HUME 1711-1776, ...), in German (LUTHER 1485-1546 to begin with, KANT, HEGEL, ...). (We leave it to the Reader to insert the names of poets especially since the authorship of SHAKESPEARE is controversial.) A more recent illustration is furnished by the Danish language, which can be said to have been finally moulded by the philosophic writings of KIERKEGAARD's (1813-1855) as much as by the *Tales* of H. C. ANDERSEN's (1805-1875) at a time when romanticism and the early form of existentialism flourished.

This linguistic phenomenon illustrates another thesis, asserting that more and deeper knowledge is by and by reached not only along, and through, the application of *reason* in order to establish *truth* (as truths, - rather than Truth in the singular which is more or less thought of as 'final', through theology), but also by the means of a 'poiesis' which implies the momentous act of *creation* which should accompany *insight* in order to achieve a whole which is both 'open' and 'closed' like a box having contents in its available space.

For, after all, thought is made of ideas requesting illustrations. Good, solid ideas have the nature of 'universals'; good, solid illustrations have the nature of (singulars or) 'particulars'. The application of reason leads to the elaboration of 'theories': as such will-be universals; this is the way 'science' in its broadest understanding progresses; we know that these theories require verification (if not falsification when they are unfit), such (falsi-, resp.) veri-fication being mediated by experience, i.e. practically, by experiments which are nothing but particulars. Philosophy proper is not just theory in this sense and does for that not reduce to a science, for it is, according to GABRIEL MARCEL, a 'second reflection', hence a reflection *on* something that already in itself involves a reflection, e.g. on science, like in the 'philosophy of science', and therefore always takes the shape of a 'philosophy of ...'; whereas *metaphysics* can (and should) be considered a sort of science proper (yet a science *sui generis*) since it has an 'object' which, not being however commensurable like in physics proper dealing with (the collection of all worldly) objects, is 'incommensurable' since it coincides with the (all embracing, infinite and perfect) 'Being of God'. Therefore, metaphysics aims at establishing 'theories', i.e., universals, which should be verified by, or say illustrated in, particulars. Theories are of course not deduced, but inferred, from experience: in this latter case the experience we accumulate about the incommensurable. The other, ordinary sciences operate in the same manner in reference to commensurable objects, even though these may be huge ones and at enormous distances from each other like in astrophysics.

Poetry is of course not a science *(sui generis),* it is an art, but an art, indeed *sui generis* too as compared with the other ordinary arts, because it aims at the creation *(poiesis)* of works assumed to be truthful illustrations of the Divine, i.e. if you please, of the in-con-ceivable (instead of the in-com-mensurable) Idea of God. For, to conceive is also to com-prehend; indeed, the Latin *capio, is, ere, cepio, captum* from which the vocables (con-)cept, ... con-ceivable ... are derived means to grasp (both concretely and/or abstractly, including what happens during sexual intercourse, ... i.e., after all, a 'creation'), and the best that can happen with (such a) conception is, that it be beautiful. Therefore, indeed, geometry is the most beautiful possible (re-)creation of the world: *κόσμος* for the Greek, according to PLATO the Order in, or according to, which God has created it. Poems are works of art; Poetry is therefore to Art what Metaphysics is to Science; yet in a reversed manner, because works of art, like here poems: *ποιήματα* as prime products of the human mind are as particulars the best possible suggestions (instead of verifications) as to what the all-embracing Idea of the Divine is "to be". All genuine works of art as created within any particular art are to be suggestions of what is implied by such and such idea. Especially, in poetic creation, the all-embracing Idea of the Divine is *the To-be,* in French *l'Etre* (with a capital *E*), in German *das Sein,* for which the Latin language fails to offer a vocable, except *Deus* (cf. any Dictionary) - in any case neither *natura,* nor above all the mediaeval *existentia* - , whereas the Greek language might use τὸ ῞Ων (which after all is God!) - yet neither εἶναι, nor just ὂν, nor πάντα τὰ ὂντα which is a plurality, nor φύσις of course. Current English has no word for it either, for it is not either Being (since this is but a participle *present,* i.e., attached to a temporal tense of the conjugation); therefore the only choice left is to make a substantive noun out of the verb - and call it *the to-be.*

The 'poem' entitled *"The To-be"* is still kept waiting for publication (and presumably for creation)! Otherwise it could (and should) have been included at this place of our text.

2. *Comment.* - Owing to the aforesaid, setting up the list of the main vocables implied in our analysis of the problematics about God, World and Time as well as related matters, becomes commendable. Owing further to the danger of ambiguity of individual languages, a threefold correspondence from English to French and to German - as the main languages of philosophical expression in modern times - may help to avoid equivocation. Furthermore, since the approach to the said problematics may be performed not just from diverging standpoints of schools, but rather according to various categorial assessments of philosophy, i.e., whether the mentioned ideas, notions and concepts refer (mainly) to e.g. metaphysics, to epistemology ... or to such and such division or chapter of philosophy, or even to general philosophy as such or - why not - to religious thought, or to positions like the philosophy of ... existence, essence, ..., or possibly to several of these, we might as well put down the possible categories to which each term may refer.

These rules will lead to a scheme which shall contain the English, the French and the German designation; as to these three languages, we should realize that *nouns* in English may be either, or often not, preceded by a definite article, or by an indefinite article, referring to three different, say intentional, significations, as e.g. when some argument concerns God (without an article), the God, or a god (with the indefinite article),

or

Time ("), the time, or a time

World ("), the world, or a world

(e.g. chosen among LEIBNIZ's 'possible worlds', or the world of ... say businessmen).

We shall therefore in our list make it clear, which of the three cases is meant. As to the categorial references, here are the main possible cases (and their abbreviations, in square brackets [ ] further down): general philosophy [g. phil], religious thought [rel.], mystique or mystical thought [myst.] not to confuse with mysticism, which is an ideological doctrin (or often the contemptuous designation for an ideology), metaphysics [m.phys.], philosophy of essence [ess.], philosophy of existence [exist.] not to confuse with existentialism which is a doctrine (or even an ideology) like socialism[1], philosophy of time [time], philosophy of science [phi. sci.] which is reducible neither to mathematics, nor to logic, philosophy of art [phi. a.] which is not reducible to æsthetics, philosophy of morals also called moral philosophy [mo. phi.] which is not reducible to ethics, morals of the self [mo. self], theory of knowledge [th. kn.] often called epistemology in English, less often in French and neither in German (*Erkenntnistheorie*) nor e.g. in Italian (where it is called *gnoseologia,* see later), philosophy (or theory) of value or axiology [axio.], logic [log.], mathematics proper [math.], æsthetics proper [æsth.], ethics proper [eth.], philosophy of technics [phi. techn.] which is not to confuse with technology which is an American neologism for technics proper, philosophy of the person [phi.

[1] An ideology makes a concept - like Society, State, Mystery or what have you - to an 'infinite' measure of all things, - which is self-contradictory, since per definition a (unit of) measure must be smaller than the measured 'things'.

pers.], physics proper [phys.], cosmology [cosm.], etc. etc. Of course one and the same notion, or one and the same concept, can refer to more than one category.

3. *The to-be, being, existence; thing(-ness), nothing(-ness), and other nouns: Comments.* - The most fundamental word thinkable of as 'void' of any *particular* connotation, is the *verb* 'to be': *être, sein* (Latin: *esse,* yet we remember that Latin dictionaries always quote the main tenses of verbs in the first - eventually the second, even sometimes third, person; in the present case: *sum, es, fui ... esse*). From *esse* comes *essentia*, in English (the) 'essence', the adjective of which would be 'essential;' hence what is attached to 'to be' belongs to essentiality: 'to-be-ness' so-to-speak, and if we are requested to translate the French noun *l'être* or the German noun *das Sein* as obtained by substantifying the verbs, we could - and should, it seems - say in English *'the to-be',* but preferably NOT 'being' (or even Being) which immediately calls for a connotation, like e.g. being (with-)in time, or in the world, or ill ... or even absent! And surely SHAKESPEARE should be cited among the very great philosophers since he had HAMLET put the 'question' 'to be or not to be ...', allowing for interpreters as well as listeners of the play to give each their own answer, but also allowing for commentators to eventually classify *Hamlet* (and indeed perhaps SHAKESPEARE himself) say among the existentialists, or another school of philosophy, say essentialism, yet at least among metaphysicians.

However surely, the 'to-be' should be confused neither with being, nor of course at all with existence. Being is already a tense, viz. a participle, and a participle present, and the presence it involves implies an explicit togetherness with otherness as well as a tacit temporal continuity (e.g. as contemporaneity or as simultaneity), which is not necessarily meant when using the plain noun *the to-be: l'être, das Sein,* which is different from *l'étant* and *das Seiende.* The to-be is the purest, and the most, *essential.* We might say this in the following negative form: Nothing is without the to-be. Unfortunately, this saying so is already tainted with nullity, firstly because using the word 'nothing' already implies that we know what a thing (and hence 'no-thing') is, secondly nothingness and absence (or negation) of the to-be are in ordinary language synonymous, although linguistically they are not, strictly speaking, for a thing is never the to-be, since it needs existence; thus we are left with an empty saying.

Already the use of the noun 'thing' involves a risk, for it does not coincide with a 'being', - or better still: a being is not necessarily a thing, hence the compound 'nothing' carries along a still greater risk. For a thing (*une chose, ein Ding,* Latin *res*) 'is within time', it is even 'a being within time', and is therefore endowed with temporality, ... and changes! Consequently nothing - or even 'nothingness' - needs not etymologically imply more than 'not being endowed with temporality', hence is not just and radically 'not-to-be': the negation of the to-be (Latin: *nihil*, French: *le néant,* German: *das Nichts* ...). We recall that HEIDEGGER invented a verb allowing him to assert: *das Nichts nichtet,* which SARTRE translated into: *le néant néantise,* implying a formidable destruction, which is not of mere material, nor even existential, but of *essential* nature.

Human beings *exist*, i.e., they are (with-)in time, and they even change, for instance they die at a certain age. Hence they partake in the to-be *and* (moreover) in existence. To exist is 'to be (with-)in time'.

Do angels exist? Human beings are unable to answer that question apodictically with either yes or no. Should we then perhaps rather ask: Are (there) angels? This latter question is usually (popularly) identified with the former one; yet as it stands, it concerns only the to-be, not the eventual existence of angels; hence as long as philosophers do not possess criteria allowing to decide about it, it admits of no answer within human *philosophic* thought (though a religion may *dogmatically* e.g. give a positive answer). For the phrase 'there is' (or 'there are'), the French use not the verb to be, but translate it with the help of the verb to have: *Il y a ...* and the German by the verb to give: *es gibt*, in both cases in the (indefinite) singular; thus the questions would read: *Y a-t-il (des anges), Gibt es (Engeln)?* Yet these questions are always (popularly, currently) understood as meaning: Do (angels) *exist?* i.e., identification of the to-be (in French *l'avoir*, in German *das Geben*) with 'existence' is tacitly made.

If we now try to transpose this discussion from the angelic to the divine 'nature', we are actually confronted with a quasi insuperable difficulty, which has been simply (voluntarily or not?) ignored throughout the centuries by just confusing existence with the to-be. One speaks indeed with the greatest ingenuousness about the 'existence' of God, eventually about the existence of gods. As long as one keeps to the consideration of several gods or deities (e.g. commenting the Greek, or another, mythology), there is no difference between this case and the case of angels, with one difficulty however, in case one or the other deity is made responsible for the creation of all what 'there is and/or exists'. But as soon as one keeps to monotheism - which is the usual thing done in our days -, the following alternative arises: *Either* (the one) God has (or had) nothing to do with the creation of Time as well as of existing being(s), the latter having been (or being) present within time in total independance from (that) God, and then (that) God is a petty god of slight power, not possessing the almightiness to dispose of World and Time and any Thing, - *Or* (that one) God is almighty, is *the* Creator of World and Time, and then the to-be is His attribute, whereas existence - even though depending upon His good-will - is the property or nature of beings which He plunges into time to make out of them 'things' (Latin: *res*, from which the noun *reality* derives), which change (in time) and eventually collapse out of their existence. In case such things are 'human beings' (or at least animal or vegetable ones), such a collapse is called their death.

The question, 'wherefrom' this almighty God 'gets' the beings to which he grants existence, is thereby not (yet) answered. The only answer which can be given to that last question is, that He tears them out of, or pulls them up from, His own Being, for, if He had to fetch them 'somewhere else', it would signify that they already were, either as mere beings or eventually already existing, and this contradicts the hypothesis that He creates them. This is one of the most striking and remarkable findings of metaphysics, for it establishes - if Divine Creation is indeed the case - and demonstrates a tie of extraordinary and indisputable intimacy between the Creator and His creatures, and allows by the way to draw some further conclusions of the utmost bearing about the world as it is indeed experienced, in particular the fact that anti-values accompany the values (the good etc.) which are actually encountered in it.[2]

4. *The to-have and further such notions.* - Since time immemorial, Man has indeed attempted at having more and more at his disposal. There is no limit to what he has undertaken to win goods, and, in contradistinction to the 'other' animal species, he became aware of the fact that such goods can be both of material and of spiritual nature - the most primitive distinction between matter and spirit being that matter has 'weight' and is apprehensible by what is soon felt as the senses (indeed, the human 'body' serving as the 'paradigm' of materiality, whereas spirit escapes bodiliness). Realizing the difference between matter and spirit was at the dawn of civilization.

Earning thus goods which he accumulates either by purchase

[2]For details, consult A. MERCIER, *The Problem of the Imperfection of a World, Itself Created by a Perfect God* (in Foundations of Physics, Vol. 22, No.2, pp. 205-219, 1992).

or by plunder, Man can be said 'to have'; and this eventually became a distinctive sign of his nature, as different from the nature of other things, including animals. One can therefore make it into a noun to designate the concept which it implies, calling it the 'to-have'.

Other species do also gather goods for their maintenance, and some do even store part of them for the sake of surviving over the hardships of cold and dry seasons, whereby a repeated struggle takes place among individuals for the sake of maintaining habitat and the like; but that does neither evolve into trade nor degenerate into war, and it does therefore not among them develop into civilization where the 'to-have' and the 'to-be' get into an entanglement which could be designated by a further verb made into another noun: the 'to-become'.

The latter should however in no case be confused with a fourth noun as also construed from a verb, viz.- the 'to-come'. For indeed, the to-come is not an attribute of mankind; much rather, it is what happens to mankind as grasped by its awareness as that which arises from the future (Latin: *futurum* as a participle of the verb *esse* - a participle or grammatical form from a verbal adjective which does not exist any more in our modern conjugation of verbs - in English, French, German ... - meaning: 'which-is-to-come', plainly too: which shall, or will, be). *The to-come is the very paradigm of what happens in Creation.* The French for it is *l'avenir* (i.e. *l'à-venir,*) the German: *die Zukunft (Zu:* to, *kunft:* from *kommen*); practically all current languages use a similar combination meaning exactly the 'to-come'.

One is thus confronted with language and more particularly with respect to a few essential *and* existential human situations, with the following, very fundamental, nouns as obtained from verbs; they are:

the to-be, the to-have, the to-come, and the to-become;

and the question arises, to determine - if feasible - whether there is something like a hierarchy, or choice and arrangement of order, or other relationship among them.

Two authors (one poet and playwright - already mentioned further above - and one metaphysician and also playwright), have tackled such question, one by setting up, and questioning then, the *alternative* between the affirmation and the negation of such (verbal) nouns (he did it for one of them only, but it can and should of course be done for all four of them), the other by *confronting* the two terms of a pair as isolated from the collection of such (verbal) nouns (he did it in his turn for one such pair only, but it can and should be done for all six of them).

The first author was of course SHAKESPEARE (shall we not call him a metaphysician too?), who had HAMLET exclaim "To be or not to be: that is the question" (Act III, Sc.I) as if there were only that question, without specifying whether to be (also?) means to exist; whereas one could as well exclaim "To have or not to have ...", or also "To come or not to come ..." - yet there, not 'HAMLET' or any other *person* is the subject of the verb, - or finally "To become, or not to become" etc. The second author was GABRIEL MARCEL writing his famous book *Etre et avoir* (1935^1, 1970^2) and germane Essays.

Yet neither did SHAKESPEARE (or any one else) put the alternative question about the three other terms, nor did MARCEL (or any one else) explicitly confront all the terms in the (five) other pairs, e.g. in French: *L'être et le devenir, L'avoir et le devenir, L'être et l'avenir, L'avoir et l'avenir, Le devenir et l'avenir.* (The English or German translations are current.)

5. *Complementarity of reason and faith. Its overcoming in mystical thought.* - The most general *notion* devoid of connotation is *Thought.*

The most general notion supplied with connotation is *Wisdom*, its systematic treatment based on *reason* is traditionally called *Philosophy*, its cultivation based on *faith* is traditionally called *Religion.*

Secular i.e., prolonged, experience shows that reason and faith cannot rule at one and the same time. Such situation of mutual exclusion is called (since NIELS BOHR discovered it for physics proper) *Complementarity.* Thus either does Thought thrive philosophically, or it thrives religiously. In both cases, the human spirit or soul gains more and more and can be said to *have.* Hence the to-have is not uniquely of material nature, it is also of spiritual nature. Therefore the to-have is opposed to the to-be within thought as well. Hence the prime problem of both (or either) philosophy and/or religion resides in the couple of 'the to-be and the to-have', even more specifically than the Shakespearean one of the 'to be or not to be'; its problematics has precisely been clearly recognized in our Century by GABRIEL MARCEL. When approached from the to-be, it is of essential nature because of the *mystery* attached to the to-be not being as such 'in time'; when approached from the to-have, it easily degenerates into an existential problematics because the to-have imposes on the to-be a restriction which is found to be at the root of Time. The alternative of these two approaches is the token of an unequalled situation in which the human mind can be caught and from which - if reached - it cannot escape but by paying the greatest spiritual tribute thinkable of, since it overcomes the very incommensurableness of the Divine; for, the only way to avoid both and escape the alternative of essentiality or existentiality is to give up the autonomy of a choice of either philosophic thought or religious thought, and *to enter* the abode of the Bearer of the Mystery and Lord of the Temple, giving up the to-have in order to be, and that is 'mystic contemplation', the etymology of which is clear as standing then 'within the temple' (Greek: *εν τεμένῳ*, where *τέμενος* means delimited space as traced by the priest, into which Evil, i.e. the anti-values, cannot enter). Entering thus the abode of God amounts - indeed - to an equivalent of 'death' (that is precisely the tribute one has to pay), from which however one gets drawn back into daily existence owing to hunger and other material needs. This kind of 'rescue' is probably the most painful *experience* which ... not the body, but ... the human soul can go through. and it amounts to a 'verification' of the theory according to which man is - say - endowed with ***soul*** as well as with ***body***. It is seldom made, and not by anyone by the way, as it seems, unless perhaps at the very moment of death, though this latter opinion is most difficult to confirm. It is Mystic 'thought', in contradistinction to philosophic or religious thought, and takes neither the form of judgments made on object(s) by reason, nor that of beliefs about the beyond as issued by faith; it resides in the imposition of a 'light' amounting to the illumination of the soul by the very Divine source where the Mystery resides, to the suspension of Time and of course of all that is connected with temporality (the past, the to-come, the to-have, the to-become ...), in short (the image of) Death implying the profoundest silence, identity of the to-be with the not-to-be, ... which the Hindu wisdom has attempted at subsuming as *advaita:* under the negation of any difference or duality, the collapse of the 'either or'. This is neither philosophic, nor religious thought, it is pure identity beyond the metaphysical, like death is beyond the physical.

Western thought has stopped short of it, based as it is on both the Greek philosophic and the Christian religious traditions, and must for that live with the complementarity of the philosophical versus the religious; and since the reciprocal exclusion of the two within complementarity is embarrassing, hampering as it does a choice that cannot be made final, thinkers ... and even the man in the street ... make their choice upon criteria of convenience, breaking one of the bonds rather than the other, yet falsely believing that it makes them free.

But they do not get free thereby, since our freedom resides in and only in the recognition and acknowledgement, even the acceptation that men are not free to choose: this *is* the dialectics of being human, owing to death that makes them mortal inspite of being *persons,* i.e., not just beings in time, but moreover beings capable of saying 'I' (i.e., speaking in the 'first person') with the consciousness that their person remains the same 'I' throughout the duration of life. The mystic premonition of that very circumstance belongs to the deepest insights men can have. For, the consciousness of our being mortal is the warrant that we can be free inspite of our irrevocable finitude.

Let us therefore start with the little Vocabulary precisely there, rather than wait till the end before mentioning it: For indeed, if there weren't *death* in the 'to-come' *our concern,* there would be no thought, no value, no conscience, no consciousness, no person ... and no mystery either for us to bear with, and *no hope to meet God* at the end.

6. *Vocabulary.* -

Death : la Mort : der Tod [Myst., Phys., M.phys.]
How comes, that death, which is something like 'the end', comes at the beginning of our list? The answer to that question is very plain indeed: If Man had no notion of a 'necessary' end, he would be able to conceive neither of beginning, nor especially of the negation of both beginning and end, i.e., of 'eternity'. Let us not forget that while, on the one hand, *thanatos* in Greek means both 'death as a *must*' and the name: Thanatos given to the 'deity' charged with the task to take away (into the 'realm of the dead') those who are at the end

of their life (like in the Middle Ages that 'shadow of a being' having the appearance of a skeleton bearing a scythe and clad in a torn veil), whereas *athanasia* on the other hand means immortality (of a being which, thus, cannot radically quit existence). The latter however, *athanasia*, cannot, from evident experience, be attributed to bodily existence and must, *eo ipso*, concern a kind of immaterial existence or say rather 'selfness', since we should like to use the attribute of existence only for beings that manifest themselves in sensible attributes.

All people of the world have, and throughout the whole history of mankind, given to that immaterial selfness a name which, of course, is that of

Soul : l'Âme : die Seele (much more than that of spirit).

Moreover, without death, there could be no understanding of a to-come as of novelty, and surely novelty is the experience we have of the to-come, - call it surprise if you please, - and that is in its own way a 'proof' of *creation*, for which it is infinitely hazardous to deny a *source* embodying it as an *Act*, and since such source must reside beyond all what has been mentioned so far, including existence, there is no other possible known such creation than

Divine Creation : la Création divine : (die) göttliche Schöpfung [m.phys., rel.]

Hence in a very stringent way, death - which no one will seriously eradicate from experience - yields a proof of God, for it is found, say, the unique - final or radical - procedure able to transcend the incommensurable. That is, it makes metaphysics possible, for it is at the same time that, which terminates physics (and science in general) as exemplified by the Second principle of thermodynamics. At the same time, it gives a nice support to the denomination of *meta*-physics for that science *sui generis* which closes the classification of all sciences.

Freedom : la Liberté : die Freiheit [essent., myst., m.phys., poetry, mo. of the self]

ROUSSEAU was wrong in saying, that Man 'is born free', for Man is born unfree, precisely owing to his obligation *to die.* And yet Man's freedom is one of the main conditions for engaging into metaphysics, for, actually, through the acknowledgment that he is not free 'to begin with', man *becomes free,* as if he had thrown his non-freedom backwards over his shoulder. This is the reason why freedom is one 'door' through which man enters the realm of metaphysics. (The two other doors are: death - already mentioned - and love - see later.)
Freedom is not a state of existence within the finitude of man; it roots in that which transcends finitude. Consequently, it belongs as an object of thought to those domains which are concerned with the incommensurable, i.e., with *metaphysics* within the pursuit of truth, with *poetry* within the creation of beauty, and with *morals of the self* within the promotion of the good. Yet owing to what has been said in the former paragraph, it belongs also to essential philosophy and to mystical thought as well.

Love : l'Amour : (die) Liebe [g. phil. thought, relig. thought, essen., myst., th. of kn., m.phys., poetry, mo. of the self]

Love is a power, as a matter of fact the greatest power there is, for it proceeds within the infinitude of the to-be, whereas e.g. mathematics, which is also a power, - while it is a fact that it works beyond the strict finitude of existence, since it conceives e.g. of numbers exceeding the number ℵ,

and reaches consequently in a certain sense the domain of incommensurability, - does however not reach the to-be of divine origin, though PLATO was certainly right when he said, that God always proceeds "geometrically" (i.e. mathematically) as the Demiurge, i.e. in His handicraft-like creative activity. In strict analogy with this, the arts rest on a mathematical-like - say architectonic - ability of aesthetic (instead of mathematical) nature when elaborating any will-be work of art, be it a symphony, a picture, ... a poem, up to works closest to divine creation, mastering the commensurable and incommensurable, whereas it needs a still greater ability to reach the to-be. Yet there, love becomes a will. And the same can be said with respect to all fields of moral nature where the 'metaphysical' (i.e. the *meta*-moral) stage is reached within morals of the self; there again, love is the power which, by mastering the most advanced ethical norms, allows to transcend all previous stages in order to reach devotedness towards the divine. Yet there, the 'power' makes room for a 'must'.

This analysis is of epistemological nature, and situates love also within the theory of knowledge, because the three cardinal values of truth, beauty and the good are approached along the three pathways of science, art and morals as helped and guided by the three powers implied in mathematics, aesthetics and ethics, the superlative of which is precisely of the nature of love. And even though it is not usual to establish such a lineage, its correctness cannot be denied and has, moreover, the advantage of calling the attention to the circumstance, that 'knowledge' reaches up to the most advanced within the domain of thought. There is no need either to insist and dwell very long upon the fact that love, - the very power indeed just nominated, - is of utmost importance within religious thought. However, there, its 'power' is found distinct: as a matter of retroactive 'complementarity', from the one invoked within philosophical thought. For, 'philosophic love' is 'selfish'. whereas 'religious love' is 'selfless', even though in both cases a 'power', whose unique and common root lies within mystic thought found to be *the* power to overcome the kind of death-like experience as described further above.

Metaphysics : la Métaphysique : (die) Metaphysik [phil., th. kn.]

has been so far quoted so often, that a further comment would be superfluous, if not to make clear, that it rules, at the level of incommensurability, the quest for the cardinal value called truth.

Since truth is one of the three cardinal values known to axiology, namely the one looked for 'objectively' by the sciences, there must, and will, be two enterprises more: for the creation of beauty as subjectively promoted by the arts, and for the pursuit of the good as established conjectively by morals, both at the said level of incommensurability; they are (with no further comments):

Poetry : la Poésie : Dichtung [phil., Ess., Th. kn.]
Morals of the Self : la Morale de soi-même : Selbst-Moral [ditto]

Yet they should - all three of them - be distinguished from the following activity (inspite of certain historical connections):

Ontology : l'Ontologie : Ontologie

(sometimes confused with, or proposed as a definition of, metaphysics), which claims to be a 'science of its own' of both being and the to-be and could as well be called the philosophy of *Essence : l'Essence : die Wesenheit*

but not the philosophy of

Existence : l'Existence : Existenz (sometimes called *Dasein*)

which we comment upon further down, since it needs first

explanations about Time, the To-be, the To-come, the To-have, the To-become.

One important circumstance which is at the very root of the possibility to gain knowledge and which should, therefore, explicitly be quoted at the beginning of the

Theory of Knowledge : Th. de la Connaissance : Erkenntnisth.
(sometimes called *Epistemology*, though the latter is less!) is that one needs *three* notions in a relation of interdependance each time one claims to take a further step that yields new insight. This condition might be designated as the *trinitarian principle.* As a matter of fact, it constitutes the foundation of epistemology. It was unconsciously used by NEWTON when founding Mechanics; in other words: at the very beginning of (theoretical) physics (called by him Natural Philosophy). But this principle is not only valid for physics, not even only for science: it should rule the foundation and development of all 'theoretical knowledge' looked for as a system of thought. It is therefore most astonishing that NEWTON did not realize the necessity to also apply such a principle at the very basis of *theological* argumentation and fought against the Idea of Trinity as placed by JESUS at the very basis of His teaching. All those - very often scientists as influenced by ISAAC NEWTON - who refuse to accept the idea of the 'Holy' Trinity, make thus an epistemological mistake.

In the following, we indicate a number of such triads which make sense, - very good sense indeed, - i.e. from which *sense* can be implicitely given to each of the three terms they implicate, whereas when standing alone, such a term remains dismissed of all significance allowing it to be used in a clear manner within cognitive progress. (The Reader will by himself find the various attributions to [...]):

The To-Be : l'Être : das Sein (not to confuse with Being),
The To-Come : l'Avenir : die Zukunft (not the To-Become),
The To-Have : l'Avoir : das Haben.

Being : l'Étant : das Seiende,
Becoming : le Devenir : das Werden,
Time : le Temps : die Zeit.

Essence : l'Essence : die Wesenheit,
Duration : la Durée : die Dauer,
Existence : l'Existence : die Existenz (das Dasein).

So far, the second term of each of these triads looks more or less like a go-between (yet perhaps in one way rather than in the other) from the first to the third term.

There are also triads in which, rather than working as a go-between, the first term has the tendency to split into the two last terms (without preference as which is 'leading'), like:
Wisdom : la Sagesse : (die) Weisheit
Philosophy : la Philosophie : Philosophie (not to be interpreted etymologically as the Love of wisdom, but as an activity led by the particular procedure called *reason*)
Religion : la Religion : (die) Religion

according to the trinitarian relationship of
Thought : la Pensée : das Denken
Reason : la Raison : die Vernunft
Faith : la Foi : der Glaube,

pursuing
Value : la Valeur : (den) Wert as implying
Quality : la Qualité : Qualität
Quantity : la Quantité : Menge (Quantität),

and splitting into
Truth : la Vérité : die Wahrheit
Beauty : la Beauté : die Schönheit
The Good : le Bien : das Gute,

all three counteracted by
Doubt : le Doute : der Zweifel calling for either
Certainty : la Certitude : die Gewissheit or
Belief : la Croyance : den Glauben,

often in combination (by pairs among all three) which define:
Technics : la Technique : die Technik
tested by
Responsibility : la Responsabilité : Verantwortung[3],
and authenticated by
Justice : la Justice : die Gerechtigkeit.

The above elaborated scheme might serve as the weft upon which an elaborate exposition could be built, yet it is perhaps as such suggestive enough to allow the reader to develop for himself a *Theory of Thought* within which Logic would not degenerate within the objective limitations of pure mathematics but would split into the authentic subjectivity of

[3]For a justification and complete elaboration of this subject-matter concerning the all-embracing signification of Technics and the nature of Responsibility, cf. A. MERCIER, *Responsibility, the Theory of Values and the Theory of Knowledge* (Proceedings of the 'Entretiens de Prague' of the International Institute of Philosophy on *Responsibility*, Sept.5-9, 1990, publ. by the National Academy of Sciences, Prague).

aesthetics and the genuine conjectivity of ethics as well; then, thus, *Knowledge* would neither shrink into 'science proper' (as was the narrow-minded intention of a positivism also known under the name of scientism), nor reduce to mere acquaintance with the facts (strict *épistémè*), but would follow the detailed and mysterious (sometimes even weary) path of cognition as an act (*gnôsis, dia-noia*).

However, we shall sketch our own Theory of Thought in an abridged manner. A Theory of thought is not a Theory of knowledge properly speaking. Yet it can be viewed as a generalization of such.

First of all, this will be done by establishing what may be called an

Analytico – Synthetic	*'Αναλυτικόν καὶ συνθετικόν*
Dictionary	*Λεξικόν*
to the subject – matter of	*περὶ προκειμένου*
God – World – and – Time	*'Ο Θεός – Τὸ Πᾶν – 'Ο Χρόνος*

N.B.- On each its side of *Mystic Thought,* which itself dwells on silence and appears thus as *implicit,* there are two kinds of thought: *Philosophic thought,* and *Religious thought.* They both approach the problematics of GOD-WORLD-and-TIME: the first one through the application of *Reason* (Greek: *λόγος*), the second one through the means of Confidence (or Faith, Greek: *Πίστις*).
On page 68 is a list of Some bilingual correspondences between English and Greek (or Greek and English) concepts used either as words or as pre-, resp. suf-, fixes, which shows the great variety in the use of Greek linguistic roots in modern, especially the English, languages.
On p. 69 is a scheme or *Chart,* kind of Table of information showing how by 'reason' (as conceived in its most extended and complex ramifications) thought works in its pursuit of the irreducible, so-called cardinal values of Truth, Beauty and the Good, the critical evaluation of which leads to a *synoptic exposition of philosophy* in its analytic as well as its synthetic purposefulness.
Especially, it is found that there is a transcendent approach of the cardi-

nal values within the domain of the *incommensurable*, which leads (philosophically minded) thought towards the Divine, i.e. beyond World and Time, where the Divine can be, even must be said to part into Divine truth, Divine beauty and Divine good. The steps leading there are in Greek rendered by words preceeded by the preposition *μετά* used mostly as a prefix, like in *metaphysics*.
Religious thought cannot easily be similarly reduced to a 'chart', precisely because it does not rely on reason but on faith. Yet it 'reveals' in its own specific manner the cardinal 'values', shall we say of Divine bounty, Divine beauty, Divine truth, - beginning their enumeration with the one approach mentioned last on the philosophic synopsis, because when not *λόγος*, but *πίστις* acts, the beginning is at God's, not at Man's end; yet, religion being that other human thought (beside philosophy), it involves from its very start the community of God and Man and puts the conjective before the subjective and the subjective before the objective, i.e. pure duty first, then pure will and finally pure power, and it does it at the highest level from which, by way of descendence (instead of transcendence), the usual cardinal values are recognized as coinciding, at the human level, with the ones disclosed by human undertakings. And thus a resemblance arises which overcomes the *complementarity* between philosophy and religion: the likeness of human, to what we cannot call but Divine, thought, including the establishment of truth, the creation of beauty and the promotion of the good, - the immensity of the jump from the human image to the Divine archetype being made possible by Love (only).
Yet it should be made clear that the explanations just given in this N.B. are not themselves of a specific 'philosophic' nature. Nor are they of specific religious concern. Much rather, they establish the complementarity which supplies man with a *choice* in his approach of ultimate insights.

Comments on that Chart will be found on p. 70.

Some bilingual (English : Greek and/or Greek : English) correspondences

English concept, (eventually as pre- or suffix)	Greek, (eventually several words)	English (eventually several meanings)
Sometimes historian	ἵστωρ	who knows (either is aware of or is versed in)
Sometimes history or story	ἱστορία	research, information, science, knowledge, literal meaning of Scriptures
	ἡ σοφία	ability in one art, sense, intelligence, judgment, wisdom, shrewdness, artfulness
lover of, friend of	φίλος, —η, —ον	
philology		science of language (rare: love of learning and literature)
philosophy	ἡ φιλοσοφία	love of { wisdom, knowledge
-logy	λογία	indicating the character, action, or branch of knowledge, of a person who speaks in a certain way or treats a certain subject
	παιδεία,	erudition
but not 'doctrin'	δόγμα, θεώρημα	dogma, principle, maxim
mĕta	μετα	(with), after, (sometimes: 'change')
	ἐπιστήμη } γνωσις }	knowledge { as organized, in the making
science	{ τὸ μάθημα, ἡ παιδεία	whole knowledge of a thing, teaching
the sciences	τὰ μαθήματα	
logics	ἡ λογική	(but not -logy)
new	νέος, α, ον	(in compounds:) neo-
	λογισμός	{ account, computation, reason, argument, confession
neologism		new-coined word; adoption of novel rationalistic or religious views
	λόγος	{ word, discourse, right to speak, sentence, proverb, fable, rumor, fame, treatise, subject-matter, reason, judgment, wisdom
faith	ἡ πίστις	belief, trust, confidence

PURSUIT OF	TRUTH		BEAUTY		THE GOOD	
	in the domain of the commensurable	in the domain of the incommensurable	in the domain of the commensurable	in the domain of the incommensurable	in the domain of the commensurable	in the domain of the incommensurable
Generic Name { Engl. / Greek }	(the) Science(s) τὰ μαθήματα (PLATO) τὰ φυσικά (acc. to ARISTOTLE)	Metaphysics proper τὰ μετὰ τὰ φυσικά acc. to ARISTOTLE: theology	Art (τέχνη, τὰ τεχνήματα)	Poetry τὰ μετὰ τὰ τεχνήματα	(Rectitude (morals) ἡ ὀρθότης (τὰ νόμιμα)	(Sanctitude (personal morals) (ἡ οσιότης τὰ ὅσια
On the level of *pure form*	Logics	Dialectics (PLATO)	Poietics	Θεοποιΐα	Deontics	εὐχή
On the level of *pure power* | *will* | *duty*	Mathematics	(Apo-dictics 'Απόδειξις)	AEsthetics	Θεοπτεία	Ethics	Εὐταξία
On the level of pure {theoretical / empirical} research	The Sciences esp. Physics {Theory / Experiment}	{Natural theology / Comparative religion}	The Arts esp. Music {Composition / Execution}	Theoplastics Θεοληπσία	The moral disciplines esp. τὰ νόμιμα) κρίσις Social & polit. life	humility (ἱκετεία)
On the level of their *adequation*	Verification	(Meditation Μελέτη)	Pulchrification	(AEsth. contemplation Θεοφόρησις)	'Bonification'	Happiness
On the *historical* level	Natural history	(Religious history fr. a nat. p. of view)	Cultural history	(Rel. hist. fr. an artistic p. of view)	Soc. & polit. hist.	(Rel. hist. fr. an ethical p. of v.)

Synoptik, AXIOLOGICAL CHART or System

indicating the
WORKINGS OF PHILOSOPHICAL
THOUGHT

as a
THEORY
OF
KNOWLEDGE IN THE MAKING
i.e., as
GNOSEOLOGY (a word or neologism absent from the Oxford DICTIONARY)
rather than as a
THEORY
OF
KNOWLEDGE AS ACHIEVED i.e., EPISTEMOLOGY
(accepted by the same Dictionary)

Comments

The following remarks might be added to that Chart.

1. With regard to the Idea of God, it should be mentioned that in a Platonic tradition, the uniqueness of God is rendered by using the definite article, saying *ὁ θεός*, 'the god', meaning that there is no other 'authentic' god, e.g. and in particular creator of World (the demiurge), even though atheistic philosophers might (falsely: see below under 10.) object to the use of such attribution. In contradistinction to that use, this very uniqueness is rendered in modern Western languages by dropping the article. It is true that in this volume, we do exactly the same with regard to both World and Time, suggesting that there is no other world and no other time. This is not in contradiction with LEIBNIZ' idea of possible worlds, since only one of them is assumed to have been created, and we shall in due time explain, that there is fundamentally only one 'time', which we intend to interpret then as 'the to-come'.

2. With regard to the Idea of World, it should be mentioned that current ancient Greek did not know of a 'cosmology: science of World', and that it did not use such a word, whereas it currently used *κοσμογονία* (for the creation, or origin, of World), *κοσμογραφία* (for the description of World), *κοσμοποιΐα* (again: creation of W.) and even *κοσμουργία* (for the same again).

3. With regard to the Idea of Time, the Greek differentiated between *χρόνος* (more as duration and in that sense time as the father of all things, but also as dated time, length of time or even lost of time), on the one hand, and *αἰών* as eternity, also as (long) period like age, generation, century, even life and length of life, and consequently existence, on the other hand. They had therefore the conception of a chronology in the special sense of the course of history for which they however said *χρονογραφία*, rather than -*λογία*. They had

even a third notion called *καιρός* i.e., the just measure, for sure, but more especially the precise time, the opportune or favorable moment, the occasion.

4. Contemporary authors have fallen into the habit to elaborate theories about more or less anything, calling what they thus elaborate by a name ending in -logy. Now *θεωρία* in ancient Greek, from designating the action of looking at ..., means contemplation, speculation (more than theory in our modern sense), and also deputation to an oracle; and the suffix -*λογία* is used in many words which mostly designate neither theories nor sciences, like in *ἀνθολογία* (action of arranging flowers), *ἀπολογία* (justification), *δοξολογία* (glorification [of God]), *εὐλογία* (praise), *ἠθολογία* (formation of the character), *θεολογία* (discussion upon divinity, science - indeed - of God as an object, also: Holy Scripture), *ἰατρολογία* (treatise on medicine), *ἰδιολογία* (particular consideration, personal talk), *ἱερολογία* (discourse upon holy matters), *κακολογία* (injury), *λεπτολογία* (subtle language), *μακρολογία* (prolixity), *μικρολογία* (pettiness), *ὀρθολογία* (precision of/in language), *παραλογία* (pretext, subterfuge), *τερατολογία* (talks about prodigies), *τετρα*-, *τρι*-, etc. (quadri-, tri-, etc.), -*λογία*, *φιλολογία* (love of humanities, ...) whereas e.g. psychology (as far as there was such) is called *ἡ περὶ τῆς ψυχῆς ἱστορία*, or there was a 'science' called *μετεωρολογία* (which however concerned celestial phenomena, yet not the weather), or *μυθολογία* which designated mere conversation and talk as well as mythology; and whereas *ὀστεολογία* indeed meant osteology whereas *οστολογία* was 'gathering of bones", BENTHAM's name 'posology' for mathematics (meaning etymologically the science of magnitudes), was not in use. The few -logies as sciences still called thus were *φυσιολογία* as the study of 'nature' and of the origin of things and one or two already mentioned cases which implied also other meanings, like *θεολογία*.

5. It should be clear that being in time (τὸ εἶναι), be a creature (μία φύσις: one thing) does not prevent e.g. a human being from eventually being within the abode of God (παροντος τοῦ Θεοῦ) called τό τέμενος (= *templum*, temple, an area which is sacred and into which evil does not enter), wherein Time is being suspended (ἔξω τοῦ χρόνου), then in presence of the mystery: That is mystique (τοῦτ'ἔστι ἔκτασις μυστική).

6. Another comment should be made upon the relationship between the three terms Value, Quality, and Quantity. To begin with, this relationship illustrates in a striking manner the necessity we have recognized at the root of the theory of knowledge to coin the meaning and significance of a concept by relating it with two others in a *trinitarian* implicit definition i.e., a postulate which can thus be further illustrated and made explicit in cases, the evidence of which makes the specific meaning of each term clearer and clearer. So, all differences considered, the analogy with the fundamental Newtonian postulate of mechanics about force producing change in time of momentum is striking, for the said significance will first become clear to the student of mechanics when he has imagined various kinds of force like gravitation, friction, electric and magnetic force ... just as the same significance, yet as displayed by Value, will arise in the mind of the student of axiology, when he considers various kinds of value like truth, beauty, the good ... as manifested when an attitude of objectivity, or of subjectivity, or of conjectivity is assumed by an individual in his pursuit of value, as both price and reward if you will, ποσότης and ἕξις, τὸ ποσόν and τὸ ἴδιον, quality being e.g. the nature (φύσις ... if not γένος), the grace (χάρις ... if not κομψεία), the goodness (ἀρετή, ... if not τό ἀγαθόν).

7. Yet there are both *stages* and *kinds* of the pursuit of value, and these stages and kinds are found to constitute a same pattern in the pursuit of each of the cardinal values,

consisting of two stages each: one based on the commensurability, the other based on the incommensurability of the said values, and of a series of steps between what reminds of concreteness and what reminds of abstractness.

8. The two steps attributed here to commensurability and incommensurability repectively constitute a split, the necessity of which was already felt by the ancient Greek philosophers, and led to the formal distinction mainly due to ARISTOTLE who characterized it in the distinction between 'physics' i.e., the study of Nature, and 'metaphysics' i.e., a study of that which transcends Nature. Yet for ARISTOTLE, it was on the one hand both the result of a felt necessity to distinguish a 'first' philosophy from a more common one, where 'first' in Greek is of course rendered by the adjective *πρῶτος*, but in that particular case was not thought of as would be the case of plain numbering by one followed by two, three, four etc. like with the ordinal numbers; much rather, it was meant to make of it a kind of thought: *φιλοσοφία πρώτη* taking precedence over all other kinds of (philosophical) thought (not religious thought however), and consequently, as the Middle Ages also thought, whose object is the Divine, as *ἡ περί τοῦ δαιμονίου* or even *θείων ἐπιστήμη*. From the point of view of the particular problematics which is ours, viz. of God, World, and Time, where evidently God is prime but World and Time are second, this would do if need be; yet, we require a 'reason' to do so, based on an acceptable argument, and the argument resides precisely in the difference between the commensurable and the incommensurable, for all that is bound to World along Time is commensurable ... yet with what? well, with measures as graspable by Man, the 'with' or "com-" of commensurability being the binding preposition between the 'units of measure' used (by Man, the author of thought) and the (worldly) 'Things' to be measured even though they may, like in astrophysics, be huge, for we possess e.g. beside inches and meters the concept of light-years to 'measure' very

big distances from galaxy to galaxy. However, when the problem arises to 'measure' God's proper qualifications (in the relationship quantity/quality that is characteristic of Value also attributed to God): His omnipotence, His perfection and what have you, Man fails: there arises an incommensurability which surpasses even the means of mathematics, since e.g. the transfiniteness of the number ℵ or even the so-called power of infinite continuous ensembles do not reach the incommensurability of the Divine, because the latter is always the Source and never the result of any possible act of thought and measurement. Even the most abstract consideration must use finite elements (of measure or comparison) as accessible to mankind, that is 'small enough' in order to 'measure' bigger ones. And it has been and still is the mistake of all 'ideologies' put forward by will-be metaphysicians, who proposed to use ideas of something huge (like the State, or Society ...) to 'measure' Man himself, thus falsifying these units of measure into counterfeit coinage.

All that is commensurable within the reach of the quantity/quality relationship of Value along the objective, the subjective or the conjective entreprises of the human mind constitutes, at the bottom of possible steps, the material of what are called: the Sciences, the Arts and (the) Morals respectively. And what is incommensurable constitutes the 'material' of a threefold entreprise which may be said to be of metaphysical order, as if - plainly speaking - the physical constituted the totality of what is worldly and the metaphysical the totality of what is godly in the approach Man makes of them.

9. In the proposed chart, we have chosen to restrict the use of the words physics (or also the sciences) and metaphysics in connection with the objective enterprise only. Even metaphysics is regarded here as a science *sui generis*, viz. the science of 'the incommensurable.'

10. As to the steps leading so to speak from the concreteness of things towards the highest abstraction of their comprehension, they stand in a pretty clear succession as suggested in the directory of the pursuit of the values, viz.(1) on the historical level, (2) on the level of their adæquation (in the Thomistic sense of *adæquatio intellectus et rei* basically defining knowledge), (3) on the (double) level of theory versus ἐμπειρία (badly translated as empirism: use rather the neologism 'empiry' which exists in French) yielding the Sciences proper in the pursuit of Truth, the Arts in the pursuit of Beauty, Morals in the pursuit of the Good, (4) on the level of (pure) power or of mathematics (which we thus find not to belong to the sciences proper) and (5) on the level of pure form with logics, while with respect to the pursuit of beauty and of the good, at a level of pure will we encounter æsthetics and with respect to the pursuit of the good at the corresponding level of pure duty we find ethics (both being understood *more geometrico*), while logics makes room for poietics, resp. for deontics, within the frame of commensurability. Within the frame of incommensurability, we encounter above the historical levels of religious history: Meditation, natural theology, apodictics and dialectics towards the establishment of metaphysics, æsthtetic contemplation, theoplastics, θεοπτεία, θεοποιία, towards poetry, and happiness, humility and invocation, εὐταξία and εὐχή towards Sanctitude.

These analogies on each step among the three pursuits of Value either within commensurability or within incommensurability may seem artificial and construed to the mere purpose of systematics; but they are not: they impose themselves because there is a strong parentage between them, resting upon their respective ob-, sub- and con-jectiveness.

11. As to the opportuneness of having the three columns in reference to the incommensurability of Divine matters, it should be made quite clear that even though so-called atheistic thinkers may feel inclined to contest their content as

non-sensical, there are actually no arguments in favour of atheism other than the assertion that arguments in favour of theism are disputable. But a 'proof of No-God' has never been attempted, for the simple reason that proofs have to assert some-thing - eventually some Capital thing (which is then infinitely more than a mere thing), but never have to assert no-thing. Now, since God is by definition not a thing, and only 'things' *exist* i.e., are in World and Time, even the designation of a 'Capital' thing does not reach God's essence, if there *is* God. In no case should one speak about the assumed existence of God. Yet one should not either attempt at proving God in the way mathematics (and mathematical logics) prove a theorem (some logicians have believed that this is the way one should follow!). One must attempt at proving Him analogously with the way Science, or the Sciences, primarily physics, 'prove World' ('the world') in the concordance of Theory and Empiry.

The same applies within the three pursuits of cardinal values. Therefore, whereas it does not make sense to speak about the 'existence' of God, it makes absolute sense to speak about the Truth of God, the Beauty of God, and the Goodness of God (cp. also further down).

12. That it may make sense to speak further about the Sanctity of God, is not an implication of the Chart just commented upon, owing to the complementarity reigning between Philosophy and Religion.

Chapter III. A 'Theory of Thought' as Generalization of the Theory of Knowledge

In the preceding chapter, the use of the word epistemology to designate the theory of knowledge was criticized on the ground that *épistémè* is restricted to knowledge as a fact and does not cover knowledge as an act as would be meant in the word gnoseology (which needs not be at all restricted to a knowledge of mere spiritual mysteries like in gnosticism and is the current Italian designation as *gnoseologia* for the theory of knowledge). Therefore it would be best if we could put forward a scheme covering both the act and the fact. In the most comprehensive way, that is after all what genuine authentic thought aims at. Therefore we might as well draft what will be the principles of a 'theory of thought'. As already mentioned, there are various kinds of thought, especially:

Philosophic Thought	complementary to	*Religious Thought*
using reason,		not using reason,
not resting on faith.		resting on faith.

Above, or between, both, there is
Mystique:
neither a regular philosophic undertaking,
nor a religious experience proper,
yet the 'silent contemplation' which masters above
complementarity
between Philosophy and Religion.

Let Religion remain undiscussed here.

Philosophy is according to GABRIEL MARCEL
a 'second reflection': *une réflexion seconde*
(which is different from *une seconde réflexion*),
hence a reflection upon undertakings which
already result from kind(s) of thought.
The latter can be classified according to four

(better to three plus one) *Cardinal Values*,
(which are each promoted by an epistemic
activity of its own).
There are four Cardinal Values (PLATO,
cf. LE SENNE, *Introd. à la philosophie*, 1939),
viz. three ordinary ones: Truth, Beauty,
and the Good, and one extraordinary one:
The Divine (rather than Love as e.g. LE SENNE
has it, for love is not a value, it is at one and the
same time 'the greatest power, will, and
obligation' owned, practiced, and felt by
human beings - as well as by God of
course). And as a power of such 'measure,'
it grants Man as a subject the capacity
to 'judge upon God as its Object',
making thus possible 'Metaphysics
as the Science of the incommensurable.'

The whole procedure concerning judgments is part of the theory of knowledge. For indeed, judgments must be classified in parallel with the cardinality of values, as resulting from specific *attitudes*. Judgments about truths - to begin with - require of a human thinking *subject* a so-called *objective attitude,* which obtains if, and only if, the subject makes itself the master of its object (by throwing so to speak the latter onto the plane of its operation of thought: *ob-jectum*), whereas the *subjective attitude* obtains when the subject makes itself the servant (*sub-jectus*) of its object. In these ways, objectivity leads to judgments concerning 'truth' and is the position adopted by the sciences, whereas subjectivity leads to judgments concerning beauty. In this latter case, the creating artist - and not a will-be critic of art - produces the judgment precisely by and in the created work.
So-called 'intersubjectivity' is in no case a criterium of the objectiveness of the attitude of objectivity. Especially not if it is interpreted as the 'general consensus' among, say, scien-

tists. And it must be understood here, that the greatest consensus does not obtain within science (where controversies are known to arise), but within mystique: All mystics agree. Moreover, a consensus can very well obtain among experts of art or of morals. Beside these two attitudes, there is a third one which we call conjectivity. It is prevalent when two (or more) partners interact according to the community of their interests, in other words 'with (Latin: *cum*, yielding *com-* or *con-*) one another', as if both (or all) were thrown into a togetherness where none of them has precedence over the other(s).

Finally, objectivity leads to the establishment of *scientific theories*. For truth resides in the coincidence of the sayings of a theory with the exact description of the facts of observation as concerned by that theory. In other words, and more simply: truth is the coincidence of the assumed with the factual.

Subjectivity leads to judgments concerning beauty. Beauty is promoted in works of art; an artist must submit himself to the work he intends to create in order to make it beautiful. (Many will-be artists, especially in our days, do not take that trouble!) Beauty resides in the coincidence of the intended with the produced.

As to conjectivity, it is the attitude as is necessary for the promotion of the good, which requires that it be, as a value, common to the partners involved.

Seen like that, Science and Art - when compared - are found to proceed inversely. Indeed, a theory is *a universal* (supposedly valid for all cases under given circumstances); observations under the said circumstances yield *particulars*. The Author of a theory always produces such a universal (e.g. the theory of gravitation) from (a collection of) particulars (e.g. all falling bodies). On the contrary, an artist has the idea of a universal (e.g. that of old age), and he produces a singular 'portrait' (e.g. of an old man): as a particular.

Thus science proposes, as universals, theories which are *verified:* 'made true' (POPPER's postulate being provisionally ignored) by particular(s) (facts). Art creates, as particulars, works which *pulchri-fy* universal(s) (ideas). There is in both cases a relationship of fidelity. (It is interesting to notice that in English, and in Latin too, the word true means both (the Latin) *verum* and *fidum.*)

Morals rest on judgments about the Good. KANT's categorical imperative applies there *in the sense that* the object is made *equal* to the subject: no mastery, no servitude, but complete understanding or *com*-prehension. This attitude will therefore best be called *conjective*[1].

An attitude is, formally speaking, a *mode.* Thus, there are *three modes of cognition:*

the objective mode - the subjective mode - the conjective mode

which constitute together *the modality of judgment* or judicative modality.

What has been said so far does by the way not concern only the ordinary sciences, the ordinary arts, and the ordinary branches of morals. It concerns also the corresponding approaches *sui generis* of the Divine, and this of course reveals the 'asymptotic hopelessness' of these enterprises. For indeed: How can Man master that science *sui generis,* metaphysics, by which he is expected to make God his slave? How can he master that art *sui generis,* poetry, whereby he will surrender totally to God in order to (re-)create Him? How can he master the trade by which he and God are by principle on the same level? In the Jewish tradition, God and Man talk together so to speak on a level of equality. (Jews

---

[1] This denomination, obtained by joining the conjunction *con* with *jectus* in Latin, is preferable to intersubjectivity which, besides being already used e.g. in phenomenology, does not render the obligation (or 'imperative') which is implied by morals.

are good at ethics, cf. SPINOZA, who however was excommunicated from the Synagogue.) In the Moslem tradition, 'Islam' means total surrender. (Moslems are fine poets.) In the Indo-European tradition, Man not only speculates about God ('Theory' as comprehension of the Divine is a favourite domain of thought, cf. PYTHAGORAS as well as the whole following intellectual tradition), he has approached Him, - even more than both previous mentioned peoples, - not only along a series of metaphysical steps, but also at the highest possible access of religious thought. But that compells us to jump over the complementarity, in order to reach for Religion, where Belief supplants Judgment.

What about *mystique,* then, which we have found to carry out the equivalence of the two complementary thoughts?

The answer is, that in the mystic experience, Man is no more the performer of judgment upon such and such value; much rather, he is *being judged upon* by the Lord of the Temple: in and by *con*-templation. Consequently, the three modes of the modality of judgment vanish; at the same time, belief as a modality (call it hieropoietic) is also discarded. So we encounter in mystique a third modality, that of contemplation. Yet the contemplative modality, in contradistinction to the judicative and to the hieropoietic ones, is *one,* i.e., it does not part into modes looking each for its irreducible cardinal value as happens with the judicative, or each for its irreducible cardinal practice as happens with the hieropoietic modality. (We shall not go into the details of possible modes of such a modality, like prayer, sacrifice and what have you.) Worldliness (or temporality its equivalent) is looked upon, in religion, with a disdain that is alien to the concern which the judicative modes have in World and Time.

Now:

To 'establish' truth is the business of Science, i.e., of the many 'ordinary' sciences (mathematics not included, since they do not constitute any 'science').

To 'foster' beauty is the business of Art, i.e., of the many 'ordinary' arts (aesthetics not included as one such domain). To 'promote' the good is the business of Morals, i.e., of the various domains of morals (ethics not included as one such domain).

Moreover:
Kind of Truth, kind of Beauty, kind of Good are pursued within the approach of the Divine, viz.
by Metaphysics (beyond the ordinary sciences), thanks to the *power* '*im*-plied' by Love -
by Poetry (beyond the ordinary arts: cf. J. MARITAIN's *Creative Intuition in Art and Poetry*, 2nd ed., New York 1955) thanks to the *will* '*sup*-plied' by Love -
by the Moral of the Self (beyond the moral domains within the community of existing beings) thanks to the *must* '*com*-plied' by Love.

Ordinary values can be *estimated* (units of measure ..., prices of works of art ..., number of days in prison ...).
The Divine Value is *inestimable;* thus metaphysics is a science of the incommensurable, poetry an art of the unspeakable, moral of the self a moral of the inescapable.

Hence:
Metaphysics is a science *sui generis* - Poetry is an art *sui generis* - Morals of the self are a kind of moral *sui generis*.

Finally:
To each of these three pursuits, as far as commensurability in the estimation is concerned, belongs a universal procedure itself falling into one *theoretic* and one *practic*[2] undertaking, viz.
a universal power (French: *pouvoir*) '*more geometrico*' for science, with mathematics as its theoretic and physics as its

---

[2] The Reader is requested to tolerate the use of such an obsolete adjective.

practic component,
a universal will (*vouloir*) *'more geometrico'* for art, with aesthetics as its theoretic component and composition as its practic component,

a universal duty (*devoir*) *'more geometrico'* (cf. SPINOZA), with ethics as its theoretic and law[3] as its practic component.

Hence ethics is not morals, etc., (while mystics - written in the plural for the sake of analogy - are beyond, even above, philosophy *and* religion).

Sets of 'proportions' can be alleged:
Metaphysics/Science ~ Poetry/Art ~ M.of Self/Morals, on the epistemic level,
Mathematicsc/Science ~ Aesthetics/Art ~ Ethics/Morals, on the theoretic level, etc.,
whereas Mystics cover at one go the epistemic, the theoretic and the practic levels, upon the ground that it dwells in silence, the distinction between theoretic and practic being most manifest with regards to Art and least perceptible with respect to Science.

At the end of the 20th Century, we are still in great need of (an) ethics *more geometrico*; perhaps the (mathematical) theory of sets and special logical methods will furnish 'good' starting points. Aesthetics possesses already a number of procedures *more geometrico* (in music, in dance, in drawing ...), which one learns at specialized schools.

The most conspicuous way to 'create' is to 'break' a (lower) symmetry by a (higher) symmetry (cf. *Poïesis, source et moteur de connaissance,* loc.cit.). This is perhaps the most general principle leading the pursuit of the values (including science, - as PIERRE CURIE explained years ago, - and morals).

---

[3] Here, the French would say *le droit*. However, we shall remember that there are no rights (*droits*) without duties (*devoirs*) and *vice versa*.

The branch called axiology, i.e., the theory of values, must evidently cover all values; now, knowledge is knowledge of ... what? answer: of the values, of course, especially as they are manifested in the existence of beings. Consequently, an epistemology or theory of knowledge as proposed by one philosopher and an axiology as proposed by the same philosopher shall not disagree, even though axiology is part of the theory of knowledge and not vice versa.

Philosophy is (second) reflection - by reason - upon what precedes (except religion, as already explained, owing to complementarity, and mystique, which stands 'above'), including the togetherness of the parts. By reason! Hence it has no end, since there also will be a philosophy of philosophy (... of ...). Thus philosophy - by climbing up - has the tendency to culminate: never reaching the very top however.

Religion must proceed the other way down; for it starts, where there is faith, in (and at) the top. In such descent, it will meet also the values, as facets of the supreme Value from which it originates. However, it will not be able to reestablish the same structure yet 'put up side down', for faith does not 'demolish' bit by bit what reason claims to have put together bit by bit.

There are possibly such as fancy that, religion preceding (at least in their minds) historically philosophy, the task of philosophy is to demolish by reason the kind of thought which religion has been setting up by faith in the course of former times, replacing it by a 'reasonable' building of thought; this was by the way approximately what the 'philosophers' of the so-called 'Enlightenment' 'thought' in the 18th Century; but what they actually did, was to 'believe' that philosophy takes precedence over religion (as a kind of reaction against a scholasticism which might have taught that theology - read: religion - precedes philosophy). We have already explained, that a belief is forbidden within philosophic thought proper.

Now, if previously DESCARTES did indeed react against such scholasticism in the previous century, he did not fall into the trap of the Enlightenment (he was too great a philosopher to make such a mistake), whereas e.g. VOLTAIRE, who was a scanty philosopher, did, and has for that reason done a lot of mischief in that respect. And anyhow, the problem is not, to decide, which of the two: religion or philosophy, has precedence, since they are, as has been explained further above, on one footing yet in a relationship of *complementarity*, which is totally different from precedence, just as a wave and a particle do not struggle for precedence within the physical complementarity which is their own.

Actually, philosophic thought, even though it can reach very high in its attempt to culminate, will remain somewhere at a distance from the bottom from which it started in the adulteratedness of the world. Whereas religious thought, even though it can by faith descend very far towards the impure straights of worldliness, will always hang like on a thread attached at the top.
The thread can break, i.e. faith collapse. Yet philosophy will not heal from the collapse. The climb can lead to a ridge above an abyss, i.e. reason fail in front of a chasm. Yet religion will not re-establish the lost logical consequence. This implies that metaphysics will never find a discourse sufficiently, totally adequate to its object, owing to the fact that it cannot succeed in 'reaching' and subduing the Divine Value: 'God', its 'Object'. But it implies too that religion will always give to the lay listener the impression that it fails to understand the worldliness of human situations. At the same time, these remarks call the attention upon the fact that the origin of both philosophy and religion is to be found in the circumstance, that imperfection dwells in the world. For indeed, if the world were perfect, there could not arise any problem, any question, hence any contention either. Perhaps, - by being thus perfect, - the world would

be monotonous, wearisome, tedious, boring even, lacking as it were the possibility for Man to look for reason and/or faith, no such complementary needs constituting his spiritual existence.

Yet a question arises here at once: Are such explanations as the last written ones, pieces of philosophical thought, or of religious thought? Neither, nor. They belong to *Meditation* which, - cf. the Greek *μήδομαι*, - implies a ***prudence***, meaning a preparation in either of its aspects.

We could stop here with that sketch concerning a theory of thought meant to generalize the theory of knowledge and to include into it at the same time the theory of values. However, there is one more matter, indeed belonging to epistemological considerations, which should be taken into account, although it often be neglected in the development of a theory of knowledge proper. It concerns the *nature* of values as well as the *concept* of modality of judgment (and of further modalities):
Values are not things, i.e. more explicitly not existing beings. Now, *human* beings have the particular capacity *to judge*: Precisely, judgments are made on values, and this is possible because there are both the values and the anti-values (Good and Evil, etc.), allowing for a threefold relationship between judgment, value, and its distinction from anti-value. (N.B. there are various *values*, just as there are various *(laws of) forces.*) It is not known whether non human, existing beings, e.g. animals, possess the same, or at least an analogical capacity to judge not reducible to mere instinct. A dog will bell if an intruder enters the dwelling of its masters, which is also its own dwelling. But it is not clear whether it is able to judge the trespassing as a 'bad' conduct, or whether its belling is nothing but the automatic consequence of learning to bell in such circumstances. For such a dog is a 'tame', even tamed animal. A 'wild' animal will for example hide itself in some forsaken place where to die. Elephants are said

to 'bury' their deads. About worms or molluscs, we know just nothing, not to speak about plants, ... even stones etc. Where is the sharp distinction between those who do not 'know' and those who 'know', since knowledge is 'knowledge of the values'? Must we conclude, that along with creation, the Demiurge attaches them in a way such, that 'only' exceptional beings will notice them, and belong thus to the class of the chosen or select, building then an *élite* among all existing beings? Probably there is no answer to that, not even within the frame of a 'psychology of such beings' meant to generalize animal psychology, for that means that a 'soul' is either attached to some select beings, or to all things. And this seems an unanswerable query.

Still, at the same time, it assigns to Man a responsibility towards World, - through both religious and philosophic thought, - which again is of the nature of the Incommensurable!

This last statement is of exceptional interest, for it illustrates at the highest possible level the fact that, contrarily to what is commonly opined, responsibility is not a subject-matter for mere morals. Much rather, responsibility - given its very etymology (response: answer) - necessarily presupposes the togetherness of *two* 'participants' (none of which need be morals, yet one of which often is morals) in an exchange of 'question and answer', e.g. Science and Morals as in technology or medicine, or Art and Morals as in the adornment of public spaces in towns, or Art and Science as in architecture... Indeed, only when two such enterprises of comparable nature collaborate in the pursuit of, and towards bringing harmony between, two irreducible values, or with the object of bringing together two different pursuits of one same value or aim, does the problem of responsibility arise. In the above case, two such irreducible pursuits attempt at grasping the same: the Idea of God. They are: philosophy (through *metaphysics*), and religion (through *the cult*); and since Man is by nature both a philosopher and a religious being at a time (even in case he 'believes' to be an atheist), he has to precisely assume the ***responsibility*** to conciliate his philosophic with his religious thought. And that is neither a philosophic exercise as such, nor a religious assignment proper; at the same time, it is in no way a moral task: It is the attempt at reconciling the two aspects of the complementarity within Thought, a reconciliation different from what mystical contemplation achieves, for contemplation happens in silence, whereas that responsible reconciliation has to reckon with the 'noise'[4] of contradiction.

[4] Cf. MICHEL SERRES: *Genèse*, Ed. Grasset et Fasquelle, Paris 1982.

Chapter IV. Physics as Chronogeometry: Time as four-dimensional Super-Time

From what has been said so far we can, it seems, deduce a number of statements which may in their turn call for commentaries.

1. Time - Space - Matter - Radiation

1.1 Physics investigates **the commensurable.**

This, however, is not a definition of physics, especially not an explicit definition.

It is not either exhaustive, since not only physics does investigate the commensurable, for in their own way, the other 'ordinary sciences' also investigate the commensurable. Yet physics does it most conspicuously; other sciences do it less conspicuously, the reason being twofold: On the one hand, physics has a typically experimental aspect which is not totally described by saying that it repeatedly rests on observation, for its experimental activity rests, today more than ever, and presumably to-morrow still more than to-day, on the preparation of precise circumstances within precisely limited conditions reigning between objects chosen in the world for their specific interest and a specially conceived set of fittings which equip the human performer (the experimentalist or experimental physicist) with the necessary tools to perform the experiment which has been (1) *conceived* till the utmost detail to (2) *test* an affirmation as logically deduced from a theoretical consideration concerning the qualities of the said objects, or *determine* precise quantities which stand as the evaluation of such qualities. On the other hand, since experiences of that precise nature are not mere happenings independent of any expectation on the side of the physicists, the latter must also include such, as elaborate those theoretical considerations to be submitted to either veri-fication or falsi-fication, and they are the theoreticians. No other science has driven this twofold - experimental versus theoretical - activity with an intensity by far comparable with physics, so much so, that within the profession experimentalists and theoreticians have become the performers of two clearly different, though germane, tasks, even though each must keep a sharp understanding of the doings of the other, else he would run the fatal risk to deviate from the very task of the job which he thus shares, namely to establish truth as that which is common to all specific 'truths' (Latin: *veritates*) as verified ('made true') precisely in and by the collaboration of experiment with theory. In saying that, we have on purpose used the word 'experiment' which characterizes the intentional preparation of (observational) experience as described above. Of course, not only physics strictly speaking, but also other sciences (notably astronomy which, we may very well claim, is to-day a main part of physics) deal and behave in the same manner. Yet apart from astronomy, but already from chemistry on (if we think of a list as e.g. to be found in a catalogue describing the structure of a Science Faculty at any University of the World), the (natural) sciences do not so far distinguish within their respective professions the two tendencies with a comparable sharpness. This is of course due generally speaking first to the circumstance, that they are, say, 'less advanced' than physics; but it is furthermore also due - and that calls indeed for a more serious attention from an epistemological consideration - to the fact that *physical* theories (including astronomical ones) are nowadays necessarily elaborated with the help of an 'implement' of discursive nature borrowed from mathematics: each notion as well as each concept used in such theories having a *mathematical variable* assigned to it as borrowed from one or the other chapter of mathematics, e.g. (even mostly) from geometry as well as from differential calculus, so that the total discursive procedure within physics becomes radically mathematized, meaning among other circumstances that it automatically becomes tributary of the mathematical proofs involved in the theorems demonstrated within the mathematical chapter used, and this has an exceptional influence upon the assumed validity of truths as are submitted to experimental verification, whereas e.g. biological theories are practically free from such mathematical stringency, except as far as biological objects are submitted of course to physical laws like gravitation (heaviness, ...), hydro-statics (swimming, ...), fluid dynamics (flying, ...), elasticity, material resistance, etc., which however do not concern (or seem to concern) primarily the very problematics of life (*bios*), except that the only explicit version of the latter's definition refers to the ability of living systems to resist the tendency towards decay expressed by the second principle of thermodynamics.[1] Thus, as we see, physics more than any of the other sciences investigates the strictly commensurable through the systematic use it makes of mathematics. However, and inspite of that, 'life' has - so far - never been attributed a regular mathematical *variable*, and this makes it understandable, that we make a pretty sharp difference between physics 'and the other sciences'.

1.2 Yet one notion considered by physics **and** the other sciences must be made an exception, that is **Time**. For even though time is made *the* independent variable of any physical (and even other scientific) theory, it is not, for that, 'measured': it is *numbered*, which is quite another thing. In pure mathematics, e.g. in the description of curves in geometry, there arises a dependency of variables upon one another: if the interdependency of two variables is a one-to-one dependency, one of the variables, or the other, can be chosen as 'the' independent variable. In physics, where mathematics assumes a rôle as applied mathematics, this is no more the case, for time is there *ex natura the* independent variable: time is the bearer and carrier of what is to-come,

[1] It is true that thermodynamics are often considered by physicists as a 'special' chapter which, not being concerned with specific interactions, plays if not a secondary, yet an exceptional rôle and is therefore rather neglected by physicists at large, unless one is interested by the statistical foundation which can be given of its principles.

i.e. of happening, hence: 'the to-come' itself and as such the origin of all what any existing being undergoes, quite especially of course the origin of what is commonly called 'human experience', but also of any 'thing's experience', a thing being an existing being which at any moment of time comes forth (*ex-sists*) from a former state; and thus, time is that without which no being would exist, i.e., have a course of any noticeable kind, in other words no history but also no future, even though it may 'look identical' at two following 'instants of time'. And since such identical aspect may very well be due to a difference so slight, that the latter remains practically unnoticeable to any attempt at making it the object of observation, we may postulate that 'nothing', i.e., no thing whatever remains identical in the world; hence Time is the Donor - yet the bodiless, incorporeal donor - of change. Of course, none of the 'definitions' just given is either an explicit, or an implicit definition (in HILBERT's sense). We may assume that the founder of (modern) 'natural philosophy' (i.e. physics): ISAAC NEWTON intuitively realized the impossibility to explicitly account for time and that he therefore found himself bounden to produce by postulating it a relationship of three prime notions that would express, or at least exemplify, the very nature of change, viz. $\vec{f} = d\vec{p}/dt$ (*force* = *time* derivative of *momentum*): a very precise form of postulation for which he had to *invent* (differential and integral) *calculus,* - an invention of such formidable consequence that it is comparable with what happened when at the time of PYTHAGORAS' the Greek invented *geometry*. NEWTON was certainly helped in this by the attempt made by JOH. PHILOPONOS (first half of the 6th Century A.D., exact dates unknown) suggesting what *momentum* is to comprehend, maybe through the writings of GALILEI who had read the writings of PHILOPONOS even though the latter - being monophysite - had his works put on the *Index.* Whether LEIBNIZ, who at practically the same time also invented calculus, was inspired too by the necessity to implicitly involve time in a threefold relationship, does not seem to have been the case, for he has explained that he received an inspiration towards it from passages written by BLAISE PASCAL which the latter's sister had put at his disposal to read. LEIBNIZ was not fascinated by the understanding of Time, but of World. Yet by the way, as has already been suggested, Time and World are inseparable and ought on their own to be put together into a threefold relationship analogous to what happens with NEWTON's postulate. For, there, the rôle of representing World is assumed by precisely momentum as attached to any 'thing' of (or in) the world, much better than mere mass as conceived by another pioneer of modern science: KEPLER, for whom only the motion of celestial bodies gravitating around one another attracted his attention. Therefore in a more general perspective one could and should also attempt at connecting in a threefold implicit relationship Time, World and a third notion not yet precisely mentioned, just as Force is implicitly connected with Time and Momentum. Yet World is, at least intuitively, already present in the mind as the totality of what exists, hence the third notion looked for to build such an all-comprehending and encompassing relationship must stand in some connection with Force as a sort of all-sufficing origin or 'bringing about' of World in Time, and that is - after a moment of thought - 'necessarily' found to be '(Divine) Creation', where the Divine nature of creation may for the sake of conciseness of thought be left aside in order not to give the impression that the relationship looked for is of fourfold instead or threefold structure, so we get a statement or proposition like: Time and World are the (distinct, yet common) result of Creation. And since, as we have already argued, Creation cannot after due consideration be conceived but as Divine creation, no further problem arises in that particular respect. By the way, in case the Divine nature of Creation should be considered more important or more primordial than creation itself, one could very well compare

God's Intention (in creating World with Time) as the (divine and infinite) analogue of Force (as acting within the finitude of things). One might perhaps say that Time is the superlative of impliciteness precisely because it implicates both finitude (of World) - or wordliness if you prefer - and infinitude (of God) - or transcendence if you will - since nothing is left out of the grasp which Time carries out as the To-come which it is!

1.3 At the epistemological, more precisely conceptive opposite of Time stands **Space**. That is: Space is not a notion which needs two companions to yield a sense which it gets from sharing an explanation of what is going on 'in the world', 'in time', and on account of, or owing to, some cause. Space is a construct and, as such, a remarkable testimony of our human conceptual power. It may have been conceived long before geometry was founded as the right and rigorous discourse about space. Yet even that is not quite so sure. Ancient Greek had not a precise designation for it, other than *τὸ κενόν*, i.e., as a first meaning, vacuum, another meaning going in the direction of either vanity or futility, which is certainly not the qualification which is meant to be attached to, say, a space as considered by geometry! The only word derived from it in contemporary languages would be cenotaph: the denomination of a sepulchral monument which is empty of any remains of the dead. In that sense, space is void indeed, nothingness if you please. But if space is conceived as limited and/or closed, in other words as an enclosure, it is known as *ὁ χῶρος*; even the celestial vault 'containing the celestial bodies' (*τὰ μετέωρα*) is called by that name, yet it is in a way closed precisely as a vault. Also *τόπος* can do for a kind of space, but actually it means a place, a location, eventually of a certain size or extension. Yet that is not space as such. And in Latin, even *spatium* designated some interval between points, or an extent, in any case some space as stretched, but not Space as meant by geometry. Therefore we must accept that Space as such is a concept void of existentiality, created by the human mind for the purpose of using it for computation, comparison, commensuration, concertation, configuration, connexion ... in short for the purpose of establishing formal relations between likes of abstract nature. As we have just said, it yields a testimony of the human *conceptual power,* so it is a concept, not a notion (cf. a former section in an earlier chapter), and belongs to an activity of the mind: mathematics, which is by itself a power: a power of such nature, that it is stringent but also immaterial, and this implies *non*-existence of its very elements, especially *of space:* Space has no existence of its own, it escapes temporality, it is not a thing; it could be - why not? - regarded as a being yet devoid of existence (especially since by language we express what it 'is' in phrases using the verb to be), i.e. a being in the abstract, without (but not within) the world, at most apt to help making pictures *of* the world, as if the Creator of World: God, while acting 'as a Geometer' (cf. PLATO, already quoted in such respect): conceived of Space to which a structure can - and hence has to - be attributed, i.e., such and such structure as open to the free choice of the geometer, whether human or Divine. It can therefore be, that God created a world according to some preconceived geometrical pattern as applicable also for Man to establish a (physical) theory of that world, but this signifies neither that the world is 'contained' in such a space, nor that 'it contains' the said space. We might also put it as follows: Space - especially 'pieces of space' - cannot be cut out to be sold to and bought by either amateurs or speculators, whereas 'pieces of land' can (on the Earth and perhaps one day on another planet), for these are material indeed and belong for that to Time and World, and they change, e.g. in value as calculated in money.

1.4 Now precisely, as far as **matter** is concerned, it is just the opposite of space (yet not Time itself, of course, at the opposite of which space was a moment ago said to be), for matter is meant to designate that, which is common to all what is part of, and belongs to, World. Thus we could say: World is matter, made of matter if you please; even if it changes, it remains matter. Sometimes, matter is said to fill space, or space is said to separate bulks of matter. Yet this does not change the idea, that space is somehow the void. To men, matter is all what is apprehensible by the senses. One admits, that this is also the case to animals; yet whether vegetables also apprehend it by some senses is unknown; probably yes (cp. climbing plants). However, since all existing is matter, such apprehension is the apprehension of matter by matter. Hence nothing very original is said thereby. If, starting from an hypothesis about the creation of World by God, one drops the hypothesis of God, what is left is matter. This is why a 'doctrine' called materialism amounts to the opinion that there is no God, and materialism as a posited, say even positive idea amounts to atheism as a negative idea. In such a case, the possibility to involve matter with two other notions in a threefold relationship gets lost, and one is left with a noun which says nothing even to our intuition, unless further determinations as attached to materialism are invented, leading to 'doctrines' like 'dialectical materialism' and others which fall under the criticism of being ideologies and therefore rejectable. Another possible way to follow is to develop a theory of, or about, matter. Such a theory will have to finally assume scientific aspects. For instance matter can be identified with mass as susceptible of reacting to actions of particular kinds. But then, it is already no more the 'all' and the 'nothing particular' saying as common to everything existing within World and changing with Time. Or it is thought of as the compound of 'elements', themselves susceptible of more or less precise determinations. Such elements can be conceived either as *kinds* (= natural groups of, cf. *Concise Oxford Dict.*) or as *bits* (= small pieces of anything, cf. *Same*). Older Greek philosophers thought of such kinds as earth, water, air ... or of such bits as atoms, i.e., bits that could not be broken into further bits. The latter is also the present day tendency, viz. to break matter into bits of bits until no further breaking succeeds, and then one is left with either one single kind of bits, or with several such kinds. Presently, they are assumed indeed to constitute matter thanks to reciprocal actions binding them together; they are called elementary particles, mainly reducing to very few kinds: there seem to exist electrons, protons, neutrons, neutrinos ... Yet there is no final proof that they cannot fall into further bits, call them hypothetically subparticles. The interesting, even sensational statement about them is, that precise properties can and must be attributed to them as 'the possession of' mass, spin, electric charge (of two kinds, called positive and negative since they can compensate each other), and eventually further ones as if by nature dressed up in 'charm' or what have you. It should then however be clear that as soon as one succeeds in reducing 'matter' to either kinds, or bits, or both suitably combined, there is in short *no matter* left, and physical theories (constituting in our days so-called micro-physics) which necessarily rest on the powerful use of mathematical language, indicate something akin to purposefulness in the 'material constitution' of the world, commonly 'the structure of matter', strongly suggesting that a 'demiurge' is on the look-out behind (the matter of) World, Time taking care of its changes. Or perhaps, if not even much rather, that he (the demiurge) absconds behind the to-come, loading it by and by with what may be, by those who experience it, interpreted as gifts if not rather as burdens to get rid of. Hence, after all, matter is nothing but dirt and filth, i.e., what men do not wish explicitly to see, to touch, to smell, to taste, even to hear (about), though they cannot help being compelled to do so, especially because they are themselves kind of matter, viz. human kind or simply mankind, from whose abhorrance they eventually fancy, if they trust in philosophy or more specifically in science, that they can escape owing to their intelligence in elaborating explanations - but we have stated that neither philosophy nor especially its sub-activities the sciences may rely upon trust, the latter being reserved to religious thought. Thus one is, within philosophic thought, left with the only possibility to appeal to metaphysics and its openings towards transcendence. Moreover, contemporary physics strongly suggests that matter and energy do not differ fundamentally, e.g. matter as mass is energy (up to the square of light velocity as a factor), further that energy very often comes to our awareness as radiation in waves (the 'kind' which is complementary to 'bits' as particles, especially photons). So, we can more or less sum up such considerations by saying that matter is a generic name, mostly interpreted as quintessence, for all that is perceived in time, ... including:

1.5 Radiation - especially ordinary light with its spectrum as being the special kind necessitated by, and used for, *observation* as is performed in any prepared experiment.

2. Measuring - Numbering - Light - Waves and Particles - 'Time-Space' - Reality. - Beside mere observation, experiments involve measurements. Measurements allow for three independent kinds, i.e., irreducible to one another. Each measurement is made on the basis of units of measure like one centimeter, or one gramme, or one second. (These are well-known and build the so-called c.g.s. system which is practically recognized in more or less the whole world, and used accordingly except in a few countries for some commercial purposes.) Yet actually, the number of seconds or more generally of units of time involved is never the result of a measurement proper, it is the *given* time interval for something to happen and develop till a particular change under observation attains a magnitude which is looked for, given the value of a third parameter. This is of course a consequence of NEWTON's postulate (or of an equivalent postulate, e.g. in the LAGRANGE form of analytical mechanics). Therefore the chosen unit of time like the second has a priority over all other units of measure proper, owing to time being *the* independent variable. This may give the impression that:

2.1 A watch is an instrument to 'measure' time; yet it is *not*. For actually, **time is** not measured, it is **numbered**: From some beginning, which is arbitrarily chosen (verily, it should be the 'beginning' of time or date of the beginning of creation, yet it is never, because nobody knows - not to say remembers - when creation began!): from that beginning upon which people (even peoples) agree, - called time 'zero' - time is attributed numbers: numbers of years mainly because a 'year' is easily recognizable according to the motion of the Earth around the Sun, numbers of centuries (100 years at a time), not to speak of the millennia which become so long, that human beings get into difficulty to grasp. Of course, a year containing (a little more than) 365 days, such days count easily as a more common unit, divisible, for historical reasons, in 24 hours, themselves divided each into 60 minutes of each 60 seconds. It is the very - and elementary - task of historians to proceed to the establishment of such 'numbering of time', and that is nothing but the establishment of a *calendar*. Actually, the publication of a (new) calendar is the task of such as could be called the historians of the future rather than of the past. Of course, usual: common calendars if you please, are booklets where happenings of some importance will be noted, even annotated. But in order to make them not simply more handy, but factually much more exact and at the same time *useful for the to-come,* man has invented the clock, a miniature of which is a watch (a very good 'name' for it indeed, for to watch is to envisage the future, the 'to-come'), whereas the French name for it is by far not so suggestive: *une montre* which is meant 'to show'. Still, it has by that name also the task, if not to foresee, yet to 'absorb' the said to-come and, if possible, to make it the lead for things to come and to be characterized by measures or even measurements as are made by or along physical procedures. Therefore, a clock, or a watch, is a calendar (for the to-come, strictly speaking, since it does not write down the past events on a record). The word clock itself comes from the Latin *cloca* meaning a bell, since bells were attached to such clocks at the time (approx. 13th, 14th centuries) when the first 'clocks' were manufactured; possibly, the very first clock was invented somewhere in Italy maybe by a monk who conceived of it. Such a clock is similar to a 'sun-dial', yet different from it insofar as a sun-dial is - say - a natural clock (as 'made possible' by Nature', if not by God Himself within the Creation of World, and then produced by Man): a clock however which, in contradistinction to clocks like watches, does not need to be wound up, for the motion of celestial bodies will go on - if not for ever since there is some kind of slight slowing down due to 'cosmic friction' like tides - many millennia ahead of now, especially the relative motion of Earth and Sun including the Earth's rotation, whereas clocks and watches need a steady renewal of potential energy (either in the higher position of a weight, or as the tension of a spring ...) susceptible to automatically turn into kinetic energy of its swinging piece, which cannot go by itself for ever, since the whole clock or watch is finite in size. In short, we can state that a 'unit' piece of time *cannot be deposited* e.g. at the International Bureau of Standards at Sèvres near Paris, - whereas:

2.2 A standard of length can be manufactured and kept (under constant temperature in order to prevent its lengthening or shrinking). Yet owing to such possibilities, it would be nice to know of a procedure to produce a standard length which would be independent from any physical influence, and such has been found in the **wave-length** of light as emitted by definite excited atoms. And since:

2.3 The frequency ν of a given line in the spectrum of electromagnetic waves (including of course ordinary light) and the wave length λ of the same fulfil the numerical relation $\lambda\nu = c$, c being the so-called (universal) velocity of light, a very particular relationship is found valid between such an especially chosen 'unit of length' say λ_0 and a corresponding possible unit of time ν_0^{-1} from the necessary relation $\lambda_0\nu_0 = c$. All observations made to check it have shown that c is independent of any device as prepared to measure it; moreover, the electromagnetic theory has been so well verified that the rule which it uses at its basis amounts to one of the best verified ones, viz. that c is invariant with regard to the transformations replacing one observer by another observer moving within relativity theory.

2.4 In particular, **light velocity** c (= 299 792,5±0,4 km sec^{-1} according to Int. Rad. Union. 1957) is, within present physics, declared a constant whatever be the way it shall be measured. It is not just an invariant and - as theoretical physics stand today - the most important universal constant, it is also a 'limit' which as such appears to be imposed onto both World and Time in their togetherness, hence in virtue of the trinitarian postulate to be kind of upmost gauge or standard measure to which things must conform. Thus no machinery as invented by Man (like such as conceived in ballistics) can transmit to any thing an impulse so big that this thing could move at a speed equal to, and *a fortiori* overtake, the velocity of light. Consequently, we must admit that the velocity of light is a limit as built in the process of the very Creation and that light itself is very nearly the *οὐσία πρῶτη* of ARISTOTLE'S, so-to-speak *the substance of creation* in a very strict and classical metaphysical sense: That which is permanent among changing things - its *permanence* precisely residing in its unique *velocity* from which it cannot be disconnected - and is considered a subject as implied in DESCARTES' saying (*Princ. of Philosophy,* I, 51): "When we conceive of substance, we conceive only of a thing which exists in a way such that it needs no other than itself in order to exist ... To put it bluntly, God alone is such ... But since among created things, there are those of a nature such that they cannot exist without some others, we distinguish them from such as need only God's ordinary assistance, and they are the substances ...", or as commented by SPINOZA (*Ethics,* I, Def. 3): "... quod in se est et per se concepitur ... cujus conceptus non indiget conceptu alterius rei ...", yet not conceivable without the Creation of World and Time, and moreover susceptible of transmutation into 'forms of energy' which are no more endowed with the velocity c. Light, when 'put to experimentation', reveals in the most evident possible manner the fundamental *complementarity* inherent to all what exists (but is only noticed - by men as experimental physicists - at the micro-level of things), looking either corpuscular as photon (EINSTEIN), or undulatory as wave (FRESNEL), yet never both kinds at a time (BOHR), in the second case according to a frequency ν, in the first case with a 'kinetic' energy $h\nu$, h being PLANCK's universal constant (of the nature of an action) which 'measures' so to speak in its turn that fundamental duality between particle and wave, as if beside, or rather beyond the establishment of light, the act of Creation would care for the possibility of structure. Yet, this does not exhaust the wealth of Creation. For, as we already know, three notions must be connected with one another, hence at least three universal constants can be expected to constitute the measures of World as theoretically physically approached, and if a *pure number* is to be found a kind of summary of the 'intensity of creation', a fourth constant will be necessary, else the dimensions will not combine into, and allow to compute from them, such a pure number. This means that beside c and h, some sort of magnitude must characterize the discovered electromagnetic nature of light, and one constant must embody the intensity of interaction; these will be the value e of an elementary charge found as carried by elementary particles, and the so-called dielectric constant of vacuum ϵ_0. Since four dimensions are thus involved but only three such dimensions are independent, the four mentioned constants can be combined in a way such as to obtain a pure number; that number is the so-called constant α of fine structure and is equal to $\alpha \equiv e^2/4\pi\epsilon_0\hbar c \approx 1/137$, where $\hbar \equiv h/2\pi$.

2.5 The last remarks made are important mostly for microphysics. But it would be wrong to conclude that macrophysics remains rigorously unconcerned by it. Micro- and macro- refer to 'space', yet space has been found not to possess the statute of existence. There is no fixed 'length' which would delimit the micro- from the macro-domain and be then one of the fundamental constants to be taken into account. Before the electro-magnetic interaction was discovered, only gravitation was known and involved in a physical theory as proposed by NEWTON. The only 'other' forces considered were frictions (of various nature: at the contact of gliding material surfaces or within fluids: liquids and gases), for which evaluations were proposed as possibly varying with the relative velocity of the surfaces in contact and using friction coefficients to be determined in each particular case; yet that is not a 'fundamental' law of nature like gravitation or electromagnetism. And neither the so-called centrifugal force nor the other forces of reaction are forces for which laws of nature should be looked for, they are denominations for certain mathematical terms which appear ("on the wrong side") in the equations. Gravitation was thus for a long time the only interaction known 'at the level of macrophysics'; but it acts also in case 'particles' are involved (instead of 'big bodies'), since it has been experimentally verified, that a 'jet of e.g. neutrons' or other elementary particles fall, i.e., describe above the surface of the Earth a parabola because of their weight mg attached to their mass m, g being the well-known terrestrial acceleration. Whether e.g. a neutron attracts by gravitation another neutron has not been experimentally verified, for the simple reason that such attraction is too weak to be noticed when neutrons are at visible distances apart. Other 'elementary' particles, if electrically charged, would primarily behave according to electric (and magnetic) interaction, mainly on account of COULOMB's law. It was actually by chance that COULOMB made the right guess in analogy with NEWTON's gravitation law. Actually, macrophysics proper would concern the macrocosm for which either Newtonian or Analytical (D'ALEMBERT, LAGRANGE, HAMILTON) mechanics is applicable, leading to the famous systematic canonical conjugation between coordinates and moments as the foreboding of complementarity (since the latter always applies between canonically conjugate magnitudes). Yet at the macroscopic level, the effect of complementarity is so slight, that it was not discovered until physics concentrated upon the microscopic, even sub-microscopic level. That such considerations refer to spatiality inspite of the non-existence of space is simply the consequence of applying (analytical) geometry (as invented by DESCARTES) and (differential and integral) calculus to mechanics, so that the concept of distances automatically follows as yielding a measure for the circumstance that bodies 'stand apart'. But in our days, such (big) distances are lead back to light and time, e.g. and most evidently when measured in light-years.

2.6 The most remarkable feature about the **non-reality** of an alleged spatiality **of space** is the fact, that when interpreted as such, it is found to change with time, namely to expand, i.e., not to remain identical with itself and therefore to be tributary to Time and thus to be more time than anything else: the very existence of World as a whole is temporal. Therefore, for that reason to begin with, World should not be envisaged as being space: a will-be space said to contain this and that; much rather it is *Time as a capital thing*, that thing being found to exhibit its temporality by proper means, the most contriving of which is expansion understood as giving account for the HUBBLE effect in a picture we make of it by means of geometry. Therefore the prime reason for making of World a picture based on the concept of geometrical space is not that World 'be' a space, but is that of a **four-dimensional picture of Time** which would best be called **Time-Space** (much rather than space-time, i.e., space with time glued on it): it is the correct way to represent it, not as plainly as two-dimensional pictures painted on cloth may represent life as it is experienced by physical observation of living beings, because the latter are made into suggestive, yet static things that cannot but picture an 'instant' of life, but as all-encompassing as by means of a four-dimensional time - we call it indeed time-space, or also and perhaps still better four-time or 'quadri-time' - which can be analyzed along each of the four orthogonal dimensions of a choice said to be attached to an observer possessing a watch to 'calendarize' time along its own proper time t_{pr} as well as three orthogonal 'rods' - best realized as telescopes - to 'measure' the geometrical lattice attached to it. This, in *fact* and *truth,* is what has been done through the elaboration of the theory of relativity, the three old-fashioned space-like dimensions x, y, z being glued upon the old time-like dimension of World, either the latter (alone), or the former (all three) being formally equiped with the factor, respectively the denominator, ic ($\mathrm{i} \equiv \sqrt{-1}$), c being the universal light-velocity; however, the original authors (MINKOWSKI and EINSTEIN) who formally introduced that new model of Time did not realize that this is what they did, for they were so much imbued with the idea of a real, i.e. existing, space that they proceeded expositorily the other way round, assuming even that time as a petty variable among other parameters had been made into a space-like (yet imaginary) one by multiplying it by ic, into ict. Now, this latter argument is simply epistemologically wrong. Indeed, classical mechanics consists - according to the main discovery made by NEWTON - in looking for the dependency of any (variable, i.e. changing) property of a thing existing in World upon Time as *the* unique independent physical variable (in the mathematical sense of an independent variable), e.g. especially the orthogonal 'position coordinates' x, y, z of a point-like body as measured with three rods placed along the axes of cartesian orthogonal coordinates attached to a so-called Observer. Purely formally, however, there is no mathematical difference since $(\pm i)^2 = -1$. Even more than that. When Time is correctly said to be the To-come originating in and from the future and collapsing into the past instead of supposedly running from the past towards the future, it does not make any difference to exchange intervals dt with -dt in problems of integration of the differential equations. Hence there is no formal objection against solving the latter according to the Laplacian idea based on a sort of determinism. The only objection resides against the use of the latter term: determinism, since not the future is actually determined, but only the past as that which is finished and is pure history. Now, the Theory of Relativity has dropped the very picture representing bodies possessing properties as variable in time, and replaced it by a picture where fields, nothing but fields, are assumed to depend upon the four coordinates of Time-Space (in current literature called space-time). Hence these four coordinates of Time-Space, which are a more elaborate characterization of Time (but not of space) constitute together in relativistic dynamics the complex four-dimensional, yet independent variable upon which fields with their properties, e.g. the non-collapsing components of the anti-symetric electromagnetic field $\|F_{ik}\| = \|(\vec{E}, \vec{B})_{ik}\|$ - are found to depend after due integration of the corresponding differential equations. In other words: the four-dimensional complex x_1, x_2, x_3, x_4 is not a generalization of the space position of a point-like body as fixed by $x_1 = x$, $x_2 = y$, $x_3 = z$; much rather, it is the generalization of proper time (as multiplied by ic) into (relativistic) *quadri-time*, or supertime if you wish. And if for some reason it is found desirable: either (1) to replace the pseudo-Euclidean frame of special relativity theory of the sole electromagnetism by a Riemannian frame allowing for the general relativistic description of gravitational phenomena as a substitute and amelioration of the old Newtonian description, or (2) to enlarge the dimensionality of Time from four to, say, n dimensions ($n > 4$) defining thus a more general supertime in order to allow for the completion of a unitary field theory including further interactions beside gravitation and electromagnetism, then the super-complex of the n components of coordinates in the variety thus defined is a super-time or n-time to which again we still can give the name Time-Space, but not, at all, the name space-time.

2.7 Consequently, the **Universe** (from the Latin), another physico-theoretical word for World, or *Cosmos* (from the Greek) is Time-as-structured-by-(material)-fields, and

2.8 **Reality**, or the **World of Things**, is the observable (structure of) Time-Space, according to procedures elaborated with the purpose to verify the all-embracing physical Theory as dreamed of by EINSTEIN.

2.9 From what precedes, and in that very sense, Physics can be said to have evolved into a **chronogeometry**, i.e., a regular Theory of Time-(as)-space-like (but **not** chronometry). However, we have no guarantee that present-day physics - apart from such and such improvements - has reached its final form (as was believed at the end of the 19th Century by some scientists who philosophized about the scientific achievement of their epoch and opined, that there was nothing of fundamental importance left to disclose).

Chapter V. Is World a Machine?

WHAT IS A MACHINE?

Nothing, - be it a worm, a human being ... even a motor-car, - climbs uphill without spending energy. This presupposes that, in order to overcome the resistance, it is equipped with what is called either a machine, or an engine, or more precisely with a motor, or in case it is a natural being, whether worm or human, that an equivalent organism is included in it.

A machine is an apparatus able, under given conditions. to apply mechanical power, the power itself being defined precisely as the work produced per time unit. Etymologically, a motor is more precisely a machine which imparts motion, like climbing uphill, but it has been generalized to designate any mechanical device able to impart any kind of motion, more especially rotation, whereby a system of wheels and transmission devices can transform it into linear motion of a massive body. Indeed, massive bodies are inert and do not move by themselves against resistance. Uphill inclination represents resistance owing to the weight of bodies at the surface of the Earth. As far as celestial bodies like the Sun and its planets can be said to be point masses attracting one another by gravitation and to move along trajectories as KEPLER studied them, they could be said to need no motor to entertain these motions which would go on indefinitely in absolute vacuum. But actually, they are not in an absolute vacuum, and they are extended bodies which undergo tides (not only of the water of oceans like on the Earth, but within their whole more or less viscous bodies); and this produces internal frictions at the expenses of their energy of motion; consequently, their respective motions are by and by slowed down (a loss so slight, however, that these motions go on for billions and billions of years), for there does not *exist* any motor at the disposal of the universe, which would entertain the 'life' of World, since it would have to be itself part of World.

Animals, especially human beings, are built in a way such, that they can be said to include a motor allowing them to climb again and again uphill and to do all sorts of 'other work', but again the question arises: where does the energy come from, which entertains the necessary expenditure? The answer in such cases of animal life is well known: by eating, these animals renew their energy reserve, which by the way also 'serves' to their own growth during youth, ... until they finally die.

Does a machine also die? Yes, if it is not repaired whenever necessary, and finally, it is thrown away to the car cemetery or what have you. The 'human' machine of our body is often repaired by medicine; yet in many cases, the animal machine is able to perform its own repairs, and it is not quite clear, where and how the decision to make such repairs is taken! Still, we shall all die.

If World is a machine, who repairs it? Will it after all be sent to a cemetery? The answer to the last question must be negative, since if there were such a cemetery, it would have to be part of World, which by definition leaves nothing out of its own limits.

In common language, a machine and an engine are more or less synonymous. The Oxford Concise Dictionary makes a slight difference, calling Machine an Apparatus for applying mechanical power, calling Engine a mechanical contrivance consisting of several parts. Machine is a word derived from the Latin *machina,* itself borrowed from the Greek *mēkhanē,* from *mēkhos:* contrivance, invention, device, even mere help; engine is a word borrowed from the French *engin,* itself derived from the Latin *ingenium:* nature, quality, character, spirit, intelligence, talent, genius, even cleverness (cp. *genius* in Latin: the god who is particular to each human being, from the root *gignere:* beget, create, produce).

From *mēkhos* comes *mēkhanē*: a machine (more explicitly in the days of Ancient Greece a war machine or a machine as used on the stage; or also a machination or plot, a means to achieve some purpose). The word mechanics is derived from it. Yet, scientifically speaking, it designates within physics especially as rational mechanics the particular discipline which deals with the motions of particles and bodies owing to forces.

WHAT IS AN INVENTION?

Mechanics, more recently also called dynamics, is however not the physical theory adapted to the description of the working of what are today called machines proper, because it developed in the ignorance of the fact that only a part of the energy which a machine again and again absorbs is transformed into work as is necessary to overcome the resistance which is opposed to the operation in view: for, a part - usually expressed in percentage - of the absorbed energy (as heat) is left unused for that purpose; it is namely dissipated and consequently leaves or quits the functioning of the machine, as it either simply heats its material parts and/or escapes into its environment which it also heats. It is clear that a machine or engine, being of finite size and made of solid parts which must keep at close distances from one another, must carry out cycles which again and again bring these parts into the same respective positions as formerly occupied, the reason why the transmission of work to be done requires wheels (sometimes especially cog-wheels), connected by either straps or teeth.

Now, there are no wheels in Nature[1], i.e., no 'natural wheels' in the world of things. A round, more or less cylindric, stone rolling down a hill is not a wheel; nor is e.g. (a piece of) a trunk of a tree susceptible to roll downhill owing to its own

[1] 'Nature' is not identical with 'World' as it factually is; much rather, it is World in as far as the latter is not shaped further by human inventiveness.

weight. (This is not either a machine, by the way, even though it transforms potential into kinetic energy.) The physical reason why there are no wheels in Nature is that a wheel does not simply consist of a cylindric body rolling on the ground; it must, in order to be a wheel, possess an *axis* as adaptable to a solid raised structure in a way such, that the wheel can rotate without involving that structure in its rotation. The oldest kind of such a structure was a carriage as meant to carry goods to be transported from place to place, the power necessary to transportation being spent in older times by animal, or human, 'motors or machines'. Possibly *the wheel* is *the oldest invention* made by men in prehistoric times.

The 'natural' reason why there are no wheels in Nature is its necessary connection with an axis: since while the wheel rotates, either the material axis would - if fixed to the wheel - have to get twisted and twisted until it would break, rendering e.g. in case of an organic system the arrival of sap, or blood, impossible (the 'wheel' would thus 'die' as soon as used as a wheel), or in case the axis were itself fixed to the solid structure, the wheel could receive neither sap nor blood from the said structure and would be from the beginning a dead body, i.e., unaided by any 'motor' as necessary to perform the destination of the wheel.

What there is to be found in Nature are levers more or less shaped like crow-bars; they are the limbs of living bodies: arms, legs, fingers (often, even mostly made of two or more 'levers'), and their *articulations* which allow the precise function of levers, where the destructive twisting is replaced by mere bending allowing for elastic connections, a consequence of which is, however, that motion cannot be made quasi-continuous, but has to be divided into *steps*, as is evident in marching. Such levers also perform or transmit work, requiring energy expenditure and hence revitalization just as was the case of regular motors as machines.

Moreover, once standing still, either a machine, or a system of levers (like in living bodies), necessitates a decision of being started anew: a 'will' if you please, which is never included in either the levers themselves or the machine itself, but originates from a 'center' which Science has not (yet?) been able to designate, name, or especially 'explain', not only in such mysterious cases like human decision making, but down to mere animal action or even vegetal functions like the climbing and twisting and wrapping up of creeping plants like vine etc., except that such functions are said to be included in the molecular codes of certain cells. But that does neither explain the intentionality, nor the decision making, especially to start (or also to stop) an action which, in itself, always needs a motor that assumes the energy expenditure, to say nothing about a motive as inducing an existent being to act in a particular way (eventually among a choice of possible acts, but often too simply to act purposefully); whereas what has by physiology been more or less explained is the way how ready made decisions are transmitted from the center to the driving organs along the neuromata-chains of nerves.

Somehow, even though all 'abstract' thinking is in a way or other supported by 'concrete' structures, such questions as concerning decision making and hence the creation of states of affairs by animated beings, remind so much - *toutes choses égales d'ailleurs* (especially taking the difference between Infinitude and Finitude into account) - of (divine) creation, that one is tempted to assume, that the Demiurge has endowed certain categories of beings as created by Him in the world with analogical powers of thought resembling His own power, possibly at various levels of capacity, human beings having in such respect inherited the power of thought which has been, further above, analyzed and found to part into two complementary powers covered as they are by a third one possessing an access to its mysterious origin.

What Is A Discovery?

Coming back to the question, whether World is a machine or perhaps animated by a machine-like device having its own source of energy, we must once more insist upon the fundamental difference there is between a physical system as left to the interactions reigning between its constituents and precisely an engine as capable of meeting and even resisting and/or overcoming the natural consequences of such interactions.

The natural consequences of interactions acting between the constituent parts of a system are studied by chapters of physics classified as dynamics according to the various kinds of interaction - or 'force' (Greek: $\delta\acute{\upsilon}\nu\alpha\mu\iota\varsigma$) - found to prevail within World: gravitational, electromagnetic (separable within limits into electric and magnetic), so-called weak, and strong, interaction between elementary particles, ... and perhaps further ones still to be discovered.

This last word used: 'discovered', is most important owing to its signification as distinct from 'invented'. Discoveries all rest upon 'the existence, in Nature,' of such interactions between the constituents of World. Fire, for instance, is the result of molecular re-combination of chemical elements into composites like CO_2 (as typical example of burning); and such combination rests mainly on the electric force or interaction as envisaged in the molecular generalisation of the theory developed by BOHR in his explanation of the stability of atoms. *Lighting fire* is presumably *the oldest discovery* made by man in prehistoric times.

Thus, whereas inventions consist in the production of 'things' that are not naturally present in World, discoveries consist in the proposal of means to artificially re-produce or at least imitate what is going on or simply readily available for use in Nature. The going-on is not a thing, it is a phenomenon (τό φαινόμενον: that which is noticed as *going on* in Nature); the going on is typically temporal. Discovery is therefore the act of isolating the elements, or constitutive units, of temporality. If we take this literally, the only two such elements that have been isolated *by thought* up to the 17th, resp. to the 19th centuries were gravitation (by NEWTON, with GALILEI as a precursor), and electromagnetism (COULOMB 1736-1806, OERSTED and BIOT-SAVART 1820, FARADAY 1791-1867). Of course, this does not mean that ARCHIMEDES (287-212 B.C.), who is said to have conceived of a symbolic way to designate big numbers, to have invented the crow-bar, the cog-wheel and the perpetual screw as well as their combination as called after him, when he - while bathing - exclaimed "eureka!" (εὕρηκα, from εὑρίσκω: to find by chance), had made no discovery; on the contrary, he had caught the point about two forces of the same kind compensating each other: the weight of his own body and the weight of the displaced water, both of gravitational origin (yet he did not know of their gravitational origin!); so: that was hydrostatics, not dynamics involving temporal change (of position), but simply the absence of such change. And this is a paradigm of those cases of discoveries which do not concern the very prime, formal elements of temporality, but some derived aspect of such, statics being in that particular instance an exceptional or limiting case of dynamics.

EACH PHYSICAL THEORY AN IMPLICIT DEFINITION, OF ITS OWN, OF TIME

We know already that the fundamental postulate "Force equals Time-derivative of Momentum" is an implicit definition of Time, Force and Momentum. If a particular law of force, e.g. gravitation, is proposed, a 'particular aspect of time' will ensue. Therefore, one could also say that one particular dynamics (like gravito-dynamics, or electro-dynamics

...) or theory as meant to describe by calculation the effects of one particular interaction (one particular - law of - *force*: *δύναμις* in old-fashioned speach like gravitation or electromagnetism) takes the *form* of a 'theory of time' (yet of a special and elementary way for time to unfold by revealing the specific nature of that particular interaction). In this way, a variety of 'times' can be proposed for consideration, and the times thus defined allow each to make watches. When different watches based on different such times are experimentally compared, they are found to be simultaneously regular. Moreover, the phenomena, especially the motions, they describe, can be reversed, since introducing the minus-sign before both $d\vec{p}$ and dt leaves the postulate unchanged. Therefore, the time thus envisaged can be called ***reversible.***

IRREVERSIBILITY

However, a lot of happenings are not (in any case) the direct manifestations of such elementary ***physical interactions,*** the ***causes*** they imply not being accessible to such plain laws of force to be introduced in the postulate. And they may imply consequences which cannot be made good, sometimes reaching beyond all bearing for the human soul. If they are to be considered within physics, a kind of 'dynamics' ***alien*** to the one envisaged so far must be thought of: That is ***thermodynamics,*** and the Time it implies is '***Irreversible*** Time.' For instance, dryness of climate, or its opposite, tempest, leading both to cataclysms, can be explained by physics, yet not simply from such dynamics as considered so far, but from that other 'kind' of dynamics called thermodynamics, *θέρμη* meaning heat, which is not the name for an interaction or law of force, but for a form of energy, as we know since J. ROB. V. MAYER explained it in 1842, and even as a matter of fact already since SADI CARNOT implied it in his foundation of thermodynamics (1824). The name thermodynamics for that branch of physics is perhaps ill-chosen, precisely because it is not a 'dynamics' properly speaking: it has nothing to do with force or interaction; it concerns Time during which exchanges between *heat* and *mechanical energy* take place.

In particular, the insight, that a machine necessarily carries out its tasks in successive cycles involving such exchanges, has been by CARNOT expressed by saying that in each cycle such and such amount of heat is borrowed from a (supposedly inexhaustible) source, part of which (hopefully for a good engine the technically greatest possible part) being transformed into useful work as meant to perform its due part of the task the machine is set to achieve, the remaining of the borrowed heat being expelled as not utilizable on account of the established fact, that Nature does not allow for the total use of the expenditure, there always being an inevitable loss which cannot by any means be reduced to zero. That lost heat is not recoverable, and this accounts for irreversibility. 'Good' machines lose a minimum, bad working engines lose a maximum of unrecoverable heat per cycle. The best one can hope from such an engine is realized in the case, where it performs a so-called CARNOT-cycle, depending upon the extreme temperatures between which the engine is working, and consisting of one isothermic transformation at constant higher temperature during which heat is absorbed, one isentropic transformation leading to a lower temperature, one isothermic transformation during which heat is lost, and a fourth (isentropic) one leading back to the original value of both temperature and entropy, and heat (less than originally absorbed) is lost. Other cycles are less efficient.

WORLD IS NOT A SINGULAR MACHINE

World as a whole does not undergo such cycles. It is only within the world that machines - either natural as are realized in certain living bodies, or artificially man-made - perform cycles; and these machines can only borrow the necessary heat per cycle from natural or artificially sustained sources which are within the world.

Thus, World is not a machine, not an engine, nor a motor. Yet, it has been created such, that machines can be produced in it: naturally (life), or artificially (as invented).

The Eschaton

Thus World may perhaps be said to be a complex of several machines (beside other things), say even many of them, e.g. all the 'animals' living in it; yet each machine, if it is to work as such, must absorb energy, only part of which is by the machine transformed into work done to some purpose, and the irreversibility of the processes they undergo is unavoidably connected in the long run with their decay.

Especially, World must be considered as an isolated system, since no other world is known to exist, with which it could exchange heat. Only parts of World can exchange heat among one another, and they indeed do so, e.g. by radiation, since the vacuum does not oppose any hindrance to it. More specifically, the only system which can be said to be rigorously isolated is World as a whole. But that system is not 'today' in the final state of equilibrium which it will not attain before its entropy will have reached its stationary maximum value, i.e., *its death.* This would be its natural *eschaton.*

Nothingness

And then we cannot say anything about Time, especially, if Time is the To-come, whether there should be a future behind the eschaton, unless the Creator of everything, so to speak weary of demiurgic activity, retires from it and cuts off the To-come, leaving World destitute of existence. That is, as it seems: Nothingness.

Chapter VI. Time, World, and Existence

Thus, temporality has two aspects, as far as it concerns things, the totality of which constitutes Reality: One is their Existence; the Other the loss of existence: their Death.

The latter, death, presupposes existence, whereas existence does not presupposes death; what it presupposes is Being.

Yet a precise thing, when dying as that precise thing, becomes another thing. For instance, if an animal which is alive, loses its life, it becomes a corpse. The corpse as a thing is not the same thing as the animal, for it cannot be re-animated. All things, that are beings in time (hence endowed with existence), shall lose their characteristic of being such things but not other things, and they lose it mostly bit by bit, so that in many cases they can be repaired and hence they can be said to remain, - at least more or less, - the same things. For instance an animal loses its hair, a plant loses its blades ..., but hair and blades shoot up again; eventually, either a broken piece of a thing is replaced, or mended. A lot of things which decay are restored to sound conditions either 'by nature' or 'artificially'. A temple is maintained as long as the faithful keep it in good shape. It can even be embellished, yet it does not 'grow' by itself, unless one considers those who entertain the temple as parts of the latter, constituting all together, as a community, a bigger thing having a life of its own. Yet many temples as we encounter them today are only ruins of temples of old. Even mountains fall in ruins, from earthquakes, erosion or what have you. And people - human beings - can become ruins of what they were. How long does a temple remain the same temple before being nothing but a ruin? The Parthenon was the Temple of the 'Virgin' (ἡ παρθένος) ATHENA; is the Parthenon today still the temple of the 'same' virgin?

Thus things inevitably change, and when they are left to themselves, their changing leaves sooner or later nothing but ruin (this is perhaps even more than what the Second principle of thermodynamics plainly teaches), and there is finally nothing ... more left from what they were. And since this concerns everything in the world, World itself is also doomed to become a big ruin with no one to consider it as such anymore.

The resurrection of a dead thing to its original, identically original state of the thing it was, would be the paradigm of a miracle. Miracles are radically unknown as *natural* phenomena, unless it be the very Creation of the world itself! And even the very fact that World (and Time) might have been created can be questioned by many! But then, the world's existence is radically *un-reasonable,* meaning that one cannot argue by reason about its existence, nor about the 'phenomenon' called Time. Yet, it could happen that, along the to-come, a 'thing', i.e., a being endowed with existence, is according to all possible criteria found identical with a former one without its having artificially, i.e., by human intervention, been copied on the original as a forgery. A twin might be held for an example of such. Perhaps, but then at most, all elementary particles of a given kind (e.g. all electrons, or all protons) would be examples of identical, undistinguishable things. Does this mean that Nature makes forgeries? Presumably any scientist would answer: no. Does Nature allow for resurrections? Presumably any scientist would answer: no.

The more complex be a thing appearing to the scientist to be identical with another one (which may or not have decayed), the more probable be, to him, the possibility to notice after due examination a slight difference of constitution between the two. There are not in all forests of the world two identical firs, one of which would be the reproduction of the other. There is not a star as identical with another star. There are many stars. which last enormously long time; there are many trees, which do not last that long, ... there are a tremendously big number of elementary particles: protons and electrons are very stable, lasting mostly a very, very long time. Photons are also available in incredible quantities, interacting with the former, they may produce positrons; neutrons do also come into existence. So do then nuclei, atoms, molecules ... crystals, and so on and so on, but these are assemblages of the former ones. And in a way, every *kind* of existing beings which have received a name according to a precise scientific definition is represented in World by a stately number of individuals. However no one individual of a given kind is ever considered the rebirth, i.e. the resurrection 'after death' of a given individual which would have died previously.

What will World look like, when it shall have reached its maximum entropy? Will there be nothing left but baryons (protons and neutrons) and leptons (electrons, positrons and neutrinos) in a state of such inertia (from *Latin:* incapacity) that they will be unable to interact 'any more'? Will there be among them 'corpses' left of compounds other than just elementary particles, themselves unable to decay again into simpler compounds or finally into elementary particles, owing to their stability? ... Within the Theory of (either Special or General) Relativity, no one has - so far - attempted at applying both principles of thermodynamics in order to thoroughly clarify such questions; only the First one, expressing conservation of (total) energy (of 'isolated systems'), has been taken into account [1]. The only consideration developed around the second principle within Special relativity is, that one has shown that *entropy* is a so-called relativistically inva-

[1] A. S. EDDINGTON, it is true, has considered *Entropy* and its increase in isolated systems, see e.g. *New Pathways in Science,* Cambridge Univ. Press 1935, pp. 55ss..

riant magnitude (with respect to LORENTZ-transformations) and that, consequently, BOLTZMANN's constant k is a regular *universal constant*[2].

Consequently, the endowment of existence to a being can be said to be in any case a single event. The only exception thinkable of is, that the said being be not of 'worldly nature' (whatever this means), i.e., there should then be some 'other nature', for instance what is thought of by talking about divine nature, and then it cannot be totally grasped by the same kind of (scientific, if not eventually philosophic) thought as by the one which can grasp worldly nature, the nature namely which is usually simply called Nature. So, that simple Nature is that which can be grasped by the kind of thought which looks for common, shall we say ordinary, 'truth', viz. by Science, and more specifically by physics. There may be however a Divine truth attached to Divine nature; if so, its approach, and hopefully its establishment, require extraordinary means. Insofar as one attempts at approaching it by reason, one may be said to deal with metaphysics, because Divine nature exceeds worldly nature; hence metaphysics as that science, should be called theo-*logy,* just as - by the way - physics itself should be called physio-*logy;* however, theology has traditionally developed since the Middle Ages and their scholastic tradition into a hermeneutics of religion and of its teachings; and the latter, physiology has in modern terminology become the name given to the science of 'life', which is not identical with the existence attributed to any thing. Indeed, a rather important difference must be made between life and existence: there is no life which does not presupposes existence, whereas exi-

[2]See A. MERCIER, *Universelle Konstanten, Elementarteilchen und fundamentale physikalische Beziehungen* I, II & III, (ZS. für den physikalischen und chem. Unterricht, 55, 1942); further: *Constantes universelles,* C.R. Soc. Phys. & Hist.Nat., Genève, 60, 1943, p.214 and *Constantes universelles, unités naturelles et invariance relativiste,* ibid. 60, 1943, p.258 as well as *Analyse dimensionnelle et analyse géométrique ou algébrique etc.,* Helv. Phys. Acta, 17, 1944, p.242 and *Constantes universelles,* Actes, Soc. Helv. Sc. Nat., 1944, p.61.

stence does not necessarily assume life. Life is already some kind of thingness which is more than mere, say inert, thingness (we recall here, that inert, from the Latin *iners,* means unable); however, decay is in the long run more powerful than life, since a living thing shall die by decay. Nevertheless, metaphysics is not the science whose object is Life; physiology (in the contemporary sense of the word) is that science; consequently, metaphysics is not just beyond physics, but also beyond physiology as a science; thus when we define metaphysics as the science (*sui generis*) of the incommensurable, we are not wrong, but incomplete because life is, as far as we can see today, not reducible to mere mensurability, even though quite a lot is being 'measured' by physiologists. Life has a touch of the memorable, of that which passes 'into history', and hence we might say that metaphysics is also the science (again *sui generis*) of the *incommemorable,* which involves in particular, that it is *not* a historical science[3].

On the other hand, if Divine truth is looked for outside Reason, viz. through Faith, it concerns religion, that thought complementary to philosophy and more especially complementary to science. And it is not so very strange that a noun like 'religious life' be given to an activity which may be said to be complementary to 'natural life' as approached by physiology. (N.B. 'Complementarity', according to NIELS BOHR, is a word applicable to both phenomenal appearances of existence and their epistemological approach by the activity of the mind: particle versus wave, or particle mechanics versus wave mechanics.)

Now, as science in general and the various sciences in particular seem to indicate, all that is endowed with existence, yet

[3] And should therefore not be taught (exclusively in any case) within so-called Historical Faculties at Universities, but also - and perhaps mainly - within Faculties of Mathematics and (Natural) Sciences.

not necessarily with life, if not itself existing as 'elementary particles', appears as existing in some combination of such elementary particles according to *orders* which *are 'wonders'* in their own way (more especially when also endowed with life), if not miracles in the above sense; and, since it is most astonishing that mere particles should combine into wholes of more and more elaborate structures up to, say, human beings, yet often - or at least hopefully - all subjects to such and such scientific explanations, as there already are quite a number of such like nuclear, atomic, molecular, cellular, ... theory, however less and less precise or secure theories. Perhaps, even most probably, every existing being is 'made of' *simpler* existing beings, these latter being also made of such, and so on until, by analysis, one arrives at such existing beings which are 'never' themselves 'made of' still simpler beings. They are the *elementary* ones.

One should however not forget that any analysis destroys the analyzed objects, and that nothing is more difficult than putting the latter together again, as the story told about HUMPTY DUMPTY very humorously exemplifies.

Now, atoms were believed, a century or more ago, to be such elementary things, their very name amounting to the affirmation that they are indivisible. However owing, to begin with, to the discovery of radioactivity, and also to their ionization, they were later found to be divisible. We cannot therefore absolutely exclude the possibility, that those particles which are today believed to be elementary, will be found divisible in their turn; already now attempts are made in that direction; protons can in any case 'become' neutrons if treated with electrons, and electrons and positrons can issue from photons ... It will require tremendously energetic means to break them into further 'sub-particles'. We cannot even ascertain, that there must be 'final elements' of matter. However, in the present state of physical science, such utterances are very speculative.

Be it as it may, a first, or most primitive kind of question then runs as follows:

How comes, or came, that *elementary existing beings* called particles come, or came into existence? Have all of them (still existing) been existing since ever? Have some of them died before we could have noticed about them? Is their number[4] the same now as ever? Is World nothing but a fluctuating amount of elementary particles? Why are there these particles and why (not only how) do they constitute a world of existence which is not (yet) in a state of maximum entropy? What does that mean: since, or as, ever? Does World go back to a Big Bang? Then what is it that banged that big?[5]

The interrogation marks are a dangerous piece of speach, for they ask how, or why, or when ... The 'why' is the most stringent question.

The simple question: why there are existing things, already begs the very question, since it questions our very existence.

[4]Estimated in the late thirties by A. S. EDDINGTON (cf. *The Philosophy of Physical Science*, Cambridge Univ. Press 1939, p. 170) at an order of magnitude of less than 10^{80}.

[5]It is not impossible that, - eventually as a consequence of such a big bang, - beside the many galaxies which are visible in the universe, there should be throughout the universe, spread in a way still to be determined, another kind of matter called *dark*, because invisible by means at the disposal of conventional telescope astronomy. Contemporary cosmology suggests its existence on the ground that only the gravitative action of such dark matter could be made responsible for holding the system of galaxies together. What dark matter could be made of and where it is to be found is still a problem; it could consist of weakly interacting massive particles: 'wimps', many times more massive than baryons, but nearly devoid of other interaction with the latter and consequently able to traverse unnoticed any ordinary matter made of baryons.

But why there should be elementary ones among them is still more begging the question, unless we asseverate: It is God's will, for God creates World of His own initiative, and that is, or 'makes' the to-come:

Nobody will seriously contest the fact that World exists; unless perhaps, e.g. within *māyā* (Sanskrit: 'illusion'), someone either dismisses its existence as a fallacy, or questions it, but precisely then what question exactly can be put about doubted existence is most awkward, since the very author of the question must then question his own existence.

So, either one accepts the assertion of God's creation of World, or one gives it up and gives up the explanation of existence altogether. To object to God's creation of World by saying, that the question is simply carried over to the question of God's 'own' existence, is absolutely wrong, since we know (see further above), that Existence is not an attribute that can be asserted about God Himself, for He is the Author and Decision Maker of both Time and Existence.

Yet one thing should be pretty clear by now, namely that existence proceeds from the to-come, and not from the past. For the past is, if not a radical emptiness, yet nothing but an accumulation of refuse; nothing comes out of it, and consequently determinism as a doctrine is meaningless. If You will, the Past is the wasten basket containing nothing but what is finished: terminated; and 'terminism' is not *de*-terminism; it is history, with a taste of valuation, as an -ism. Especially God, - the Creator, - is not hidden in the past: Much rather, He absconds behind the To-come.

As a matter of fact, death is not opposed so much to existence conceived as the being in time, as it is opposed to *life.* Life is, as already explained, not just existence or being in time; it is a particular kind of existence restricted to, or owned by, things which possess at least two properties: one property being the capacity to engender - in certain cases without, in other cases in connection with, the collaboration of a companion (said to belong to 'the other sex') - further beings endowed with the same capacities as at least one of the said companions, - call it reproductivity, - the other property being the capacity to take initiatives towards the coming into existence and/or towards the destruction, of other things. Given these two properties, we can speak of *living beings.* In many cases, especially among living beings whose life seems to be minimalized, the initiative capacity reduces to very little and is connected to what is called mere instinct if it is to be mentioned at all, and/or reproductivity reduces also to a minimum. This is particularly the case in simple *cells.* Sometimes, one living being is found unfit for reproduction, yet as far as can be seen, this happens only with individuals (who are then considered mishaps), but not with whole kinds of living beings since in this latter case the kind would be so-to-speak born dead as doomed to disappearance.

Various kinds of living beings are known, analogous perhaps to various kinds of existing beings beginning with the kinds known among so-called elementary particles. A systematic analysis of living beings shows that they are, like all existing things, made of elements, beginning with cells. Therefore, they are as such material beings like other ones deprived of life; but their constitution appears to be such, that the typical capacities of life manifest themselves in their particular being, on the one hand within kinds in their common features, on the other hand by degrees varying from one to another individual within one given kind. Such kinds are scientifically called *species,* grouped in genera, orders, ... kingdoms.

Already the instinct, but also the initiative lead the individuals to look for their maintenance in the search for *food*, and the food they absorb yields the energy which is necessary to entertain them as ***individual machines***. This can lead to ***aggressivity*** towards other (living or non-living but existing) beings, and such aggressivity sometimes degenerates into either destructive activity as not necessary to their maintenance (e.g. crimes among human beings), or non-destructive but 'creative' activity not either necessary to their maintenance (e.g. works of art as produced by human beings). Very often probably, 'errors' are made in the display of initiatives. Certain works however are not just gratuitous, they are found necessary for the further general maintenance of the species, like nests for birds, ant-hills, hives ...

Thus, in a way, living beings - even though they can be said to have been created by God within an assumed Divine creation - resemble God the Creator in that they can take the initiative to create in their turn, and that they also can destroy, notwithstanding, and beyond, the mere tendency towards decay as expressed by the second principle of thermodynamics. However, that resemblance stops at the point where they are all subjected to decay themselves and shall die after some time has elapsed since their individual birth (and also with respect to the fact, that God is never regarded as committing 'crimes', whereas His works are - not simply beautiful - but sublime.) Therefore a characteristic feature common to all living beings is that each is 'enclosed' within an (individual) lapse of time between *birth* and *death* more stringently as it seems than by the mere submission to the Second principle of thermodynamics, as if such an (approximate) lapse of time of longevity were included in its constitution.

Of course, it does not make sense to attribute a 'birth' to

God, and the attribution of death to Him as has been imagined by certain will-be philosophers is mere imagination in the wish to either 'poetically' explain the loss of faith among people at certain periods of history - like in the decline of mythologies -, or to justify their own a- or anti-religious feelings. Such imagined 'twilights of the gods' have by the way only been put in relationship with such and such polytheisms; the same can be said about the 'birth' of deities. But mythologies, always borne within polytheism, do not fit within, they actually lie outside, the problem of an authentic 'creation of World and Time', owing to their unavoidable anthropomorphism, which authentic monotheism systematically avoids.

Human beings, among living ones, have the tendency, especially since they have made notorious progresses within scientific knowledge, but also because they have - with *dubious* successes - won a supremacy over so-called wild life in the world (which they therefore consider 'their' world), to believe that they are the most powerful beings ... in precisely the said world (though they often forget to specify this circumstance, which implies also the specification 'within the time when they will have existed along eternity'). But they may very well be mistaken. For who knows whether e.g. bacteria will not get rid of them. Furthermore, the few last billions of years have seen the surface of the Earth changing totally its geographic and geologic shape, and this will go on during the coming billions of years anew, when none of the present continents will presumably remain as they are known today. Half a century ago, scientists thought that life has started and could only start in such and such manner under conditions of low pressure and temperature as are known to be realized under, but near the surface of the oceans. Today, forms of life have been discovered in great depths of some thousands of meters at temperatures of several hundred degrees centigrade and under huge pressure along the fractures of the sub-oceanic rifts, i.e. under physical conditions totally untenable for human and other terrestrial beings. Men will not be able to master geological phenomena on (and in) the Earth and possibly other planets not simply in the millennia to come, but in periods against which even the whole 'life' or duration on Earth of the Human kind is negligible. In his inbred arrogance, Man fancies that 'he has rights' and that 'he is free'. Actually, he has *duties* and he is *not free*. His non-freedom is due to the fact that he shall - individually - die after a relatively small number of years, mostly less than a century or so. Of course, medicine and hygiene as well as other techniques try to make it cosy for him to live and easy to reckon for to-morrow. But even the computer does not master the 'to-come'. Especially, Death is a must.

In the Latin (and Greek) tradition of speech, 'to die' is expressed by saying *animam efflare* or *animam edere* (*τὴν ψυχὴν ἀφίημι*): 'to give up one's soul'. *Anima,* for soul, has served as etymological root for that which is considered the principle of life and is encountered of course in a word like 'animal'. Is that *anima* indeed the principle of life? Cells, bacteria, ... worms, insects, ... snakes, birds ...: do all these have within their being alive such a principle as could be considered analogous to what men call their own soul, being that which is 'given up' (and/or 'back', to whom, then?) (French: *rendre l'âme*) when dying? For sure, animation is the right word for that complex of faculties which are characteristic of life in contradistinction to all other existing (yet non-living) beings in Nature. Is *anima* a natural property resulting from particular assemblages of constituent parts of the bodies of the various and many individuals among species of the animal (and vegetable?) kingdoms? Surely, it is neither the brains, nor the heart. By the way, why not indeed include here the vegetable, since after all the species within that reign have reproductive powers and possess initiatives e.g. in the fact that they can organize themselves very socially (forests, meadows ...). But plants, be they seaweeds, or roses, do give up at most a smell, or a perfume, but not a soul, when dying. That would be chemistry, a chapter of physics if you please. (At school, children like to joke by saying: chemistry, that is what smells; physics, that is what bangs!)

Apparently, 'soul' is some kind of measuring device, or better not: if it is a device, it is one to be used for numbering, like a clock or watch numbering time. Yet the soul does not number time. From scantiness till magnanimity (Latin: *magna-anima*), the soul 'numbers' values: valuables, of precious possession, for which physiology has neither measure nor memory; consequently, it is the concern of metaphysics: the science of the incommensurable and the incommemorable: shapes of existence that escape even the geometrical or mathematical and give it a supernatural taste or perfume as if of divine origin.

The Primacy of Time

Time "is not"; it "makes be".

Yet, it is not the To-be as such; it is the to-be as World, endowed with existence. World may be *cosmos*, but Time conceals it behind chaos.

For Cosmos is of old. Chaos is of new. Not the reverse, as so many ancient cosmogonies will have it. For as soon as a beautiful order appears to be established by Creation so far, Time, which is the support and myrmex of creation, changes cosmos into chaos.

Without Time, there is neither old, nor new.

Without Time, there is no birth; without Time, there is no death.

Without Time, there is no one, no two, no many.

Without Time, there is no light; without Time, there is no noise,

no feeling, no notice, no message,
no thing -
no past, no present -

there is no same and no other,

no prime, no second,
no beginning, no end -
and there is neither you, nor me.

Without Time, there is no speach, nor is there any silence, especially no logos, and no faith either.

Without Time, there is no together, no apart, no distance,

no along, and no through -

there is no keep and no loss and no forsaken at all.

No surprise and no novelty - no uncertainty and no self-evidence.

Without Time, no can, no will, and no must either,

nor would there be a within or a without -
no escape, no freedom; no cosiness, no danger -

there would be neither foe, nor friend; neither war, nor peace.

No giving, no taking. No claim. No justice.

Without Time, there would not be the World.

World is creation and repeated creation, an omnitude of novelty,

which no choice will ever terminate.

What is over, is history and has no power over the To-come, nor has it any prerogative over the world of existence.

Without Time, there is no tuition, especially no in-tuition; hence Time is not an *a priori* form of intuition, whether the latter is sensible or not. Time is THE INDEPENDENT VARIABLE: given, or imposed, as you wish. But the Giver is not of this world, for if He were, he'd be visible, at most hidden around the corner, letting at least His sandal seen in tracks that Time will however efface immediately like water wiping out all traces in the sand.

Without Time, there is no science: no physics, no metaphysics, no philosophy either,
nor is there groping or trying of any kind. Hence Time is not the output of experience,
nor is it of course the outcome of reason.

Without Time, there is no have, nor is there an ever or a never;

without Time there is no thought, consequently no complementarity within thought;

much rather, Time is the locus of complementarities, whether of fact, or of act; hence,

without Time, there is no chance nor need, even no fact and no act, as well as neither punishment nor pardon to undo either fact, or act.

In time, one finds tradition, as well as ruin, upon both of which buildings of matter or/and spirit are erected, including the imaginary as well as the premonitory ones.

Unity and synthesis are looked for but, in time, none of the two is completed; they are always left unachieved.

Time as the To-come qualifies World into the present and - as the Present - quantifies it then into the past. It is therefore the bearer of Value, since value is the coincidence of quality and quantity.

History is the story of the degradation of Value into Antivalue, itself the loss of both quality, and quantity, or at least of their togetherness. Evidently, there is no history without Time, though it concerns a past that is at most the image of time 'behind' the mirror of the present, and is for that devoid of authentic existence. Hence history is a Time Lost that cannot be found again like a lost jewel to hang again and adorn the naked neck of a queen of this world. It is a kind of emptiness like Space of geometry, but the latter is filled with points that remain uncreated, whereas the former is made of, or filled with, created events that are lost for ever. For only the To-come bears with itself Existence.

Space as of geometry is no thing, it is a concept, and it contains no thing. If an infant could speak, it would qualify space as soft, round, cosy and warm like the bosom of its mother, into which the to-come could enter and hide. But that would have to be a space endowed with temporality. *World,* not having a mother, cannot speak the same language, and must produce a vocabulary made of landmarks and references yielding an *image* of Creation. This is the way conceptualization arises; thus what it yields is a *concept* of space. Hence space is a pretext to get rid of Time and consequently to rob World of Value in order to make it a *void,* an 'emptiness', possibly as an abstract structure e.g. endowed with flatness, or with curvature, or what have you, with dimensionality of such and such grade... - a procedure, which has the provisional advantage to (seemingly) achieve unity and synthesis as looked for (see further above). However, such unity and/or synthesis is 'always' soon found to be insufficient, for the plain reason that Time is 'more powerful' than space, since it is both act and fact, whereas space is neither fact nor act: it is nothing but the conceptualization of ***nothingness***, allowing, it is true, for a whole choice of such conceptualizations; say: the choice among the possibles. Yet notice that in none of the sentences beginning, further above, with the words 'Without Time', would it make sense to replace Time by the concept of space. Therefore none of the possibles of geometry (and more generally of mathematics) are possibles 'of the world', and consequently the 'power' they presuppose is pure, and perhaps unlimited, whereas the possibles bound to World must reckon with the To-come, and therefore, the power they embody is *impure* and cannot be but usurped.

Consequently, Time and Space are not interchangeable. And while Time does not *need* (the concept of) space, it may adopt it by adapting it in a kind of companionship.

There are two such extreme adaptations; perhaps there are more, though this is not evident.

One possibility is, that Time adopts space as a kind of 'speaker/ambassador', in order to remain itself concealed as Time in the fundamental secrecy of its divine origin. This would be Time as space-like: less explicitly but more simply *Time-space.* But then, it will eventually have to content itself with geometrical 'treatises' concluded with those domains that fall under the activity of the Great Geometer. And this is the interpretation which could be given to a physical theory of time in which time remains the independent variable for which e.g. four-dimensional, geometrical space is nothing but a frame of reference for the description of what seems to happen in the world as a temporal world.

The opposite possibility is, that Time be subjected to space, and hence space substituted for time as prime actor in the working of (physical) space, introducing a concept - call it *Space-time* - which ignores the primacy of Time and interprets time as a mere supplementary dimension of its original nothingness. This latter interpretation is the one proposed by H. MINKOWSKI, adopted by A. EINSTEIN and currently used since. Its result is that space, now space-time, appears as endowed with a thingness of its own which, of course, can only concern those phenomena currently treated by the particular science of physics, for instance curvature being substituted for gravitation ... and in general the various geometrical properties attributed to that space-time being substitutes for so-called physical laws of interaction.

Formally, the mathematical formalism including the description of physical phenomena can be identically the same in both cases, i.e. for Time-space and for Space-time. But when adopting the second, one must acknowledge that one has to deal with geometrical theories which are unable to deal with any of the other cases where Time was found a prime condition for the existence of the World. And then, one cannot pretend to have reduced the world to a mere physical world.

Indeed, Time is *not* suspended by such mathematical considerations. Fundamentally, Time remains the To-come which disappears into the past through the Present. Therefore, Time is the very prime among primes. In Latin, *prima* was the designation of the beginning, especially the beginning of daylight: *"Prima hora est, sol mox surget"*. Thus, inspite of an apparent contradiction, light arises from (or with) the to-come and collapses into (the darkness of) the past.

Science is - and more generally the sciences are - somehow contained between history and mathematics. Some Schools of 'Advanced Studies' have understood that and projected it into their administrative structure (*Collège de France* as the oldest such School, Princeton ...). These sciences can be said to look for universals ruling the behaviour of observable particulars, which are all aspects assumed by the world; the universals are the scientific theories. Each physical theory (NEWTON's mechanics to begin with) is too easily interpreted as a 'conceptualization' of time, i.e. the reduction of that prime of primes into a mere concept endowed with some special attributions.

Conceptualizations like these are neither final truths, nor the Truth. For they must be 'falsifiable.' Especially, they do not dispose of Time. They make of it an object: *ob-jectum* as thrown onto the plane of consideration. In this sense, even though they do not and never actually amount to a final mastery of time, they are in their own way attempts at mastering it with more or less success. Especially, they never succeed in suspending Time, which would be the climax of that mastery.

In other words: truth, as verifiable, and consequently as falsifiable too, never gets rid of Time in an attempt to substitute itself to it.

A radically different, in its own way opposite attempt at a mastery over Time consists in substituting for actual (Divine) creation an *arti*-ficial creation of things out of what is already at hand in the world, viz. in the form of *works* said to be *of art*, doing this according to ideas held to be universally valid, and thus 'create' within the world particulars assumed to be faithful illustrations of such universals.

In a classistic tradition, such works had to be 'copies' of the beauties of the world; especially, if a mythology was at hand, reproductions of act and fact attributed to divinities; e.g. the Venus of Milo would be the most exquisite representation of womanly beauty. Yet, this was a manifestation of anthropomorphism. Later, however, God Himself would be conceived as the Beautiful:

'God is beautiful and He loves beauty.'

More recently, the very creation of particulars, often with moral content, assumed to be the crude expressions of universals, are considered works of art; they illustrate, even more than 'copies of World' in the classistic tradition, the attempt to replace the authentic creativity implied in the to-come by an artificial one which, however, instead of being objective as through the sciences, is *subjective* through the arts. Hence art is not, any more than science, a mastery over Time. Much rather, like the sciences, the arts yield substitutes for the authentic workings of Time.

There is a third possibility to establish a substitute for the workings of Time: by morals, on the conjective plane, looking for the promotion of the good. Like the two former cases, morals does not either master temporality.

*

Yet Time can, indeed, be suspended by Love only, because Love, when authentic, is a relationship: *the authentic relationship* between living beings and God Himself.

Chapter VII. The Reconstruction of Space-Time as Time-Space,

or
The To-Come where the God absconds

This chapter will be a sort of recapitulation of the former preceeding texts, yet without following the same order of matters. Much rather, it will attempt at presenting the matter both from a historico-cultural point of view and from the impact it has received from, and has in its turn given on, science, especially physics in the broadest sense, i.e. as a Theory of World.

This will lead so to speak automatically, and again, to the great question - perhaps the greatest question which can be put - about the origin of World, in others words to the problem of its possible Creation as well as to the latter's possible Author, asking both about the *when* and about the *where* of such a creation as well as *why* its Author should abscond while proceeding in such creation.. The conclusion to which we shall arrive will be that the 'where' looked for is identical with the 'when', answering at the same stroke the question implied in the kind of definition as rendered in Latin by the phrase *deus absconditus creator mundis* (better than just *mundi*).

It should be mentioned here that the text which follows is an English version, as adapted to our present purpose, of the Closing Lecture delivered in French at the Meeting of the International Association of French Speaking Philosophers at Athens already in the year 1986[1]. The reader may be taken aback by the locution ... *the God,* in French: *le Dieu*

---

[1]Published in *L'Avenir,* (Actes du XXIe Congrès de l'Association des Sociétés de Philosophie de Langue Française, Librairie philosophique J.Vrin, Paris 1987, pp. 38-54) under the title: *L'À-venir, où se cache le Dieu.*

used here, rather than dropping the definite article. In a way, this could be said to be borrowed from the (archaic) speech as used e.g. in Hebraic: *Elohim* or *Eloah* for God, or in Arabic: *Allah*, a contraction from *al-ilah,*) (*al* the + *ilah* god); or also inspired by PLATO saying ὁ θεός when talking about the demiurge: creator of the world. At the same time, it amounts to the statement, that there is only the one: unique and same god as Creator of World, even though He may be invoked by one, or by an other, historical religion; for the latter would lead to 'another world', other than 'ours'. i.e. than the one we live in, if created by another god; and this amounts to asserting, that we are concerned neither by a plurality of worlds (which would not have 'anything' in common), nor by a plurality of gods. So it may be, that *the* world we belong to is the best of all possible worlds (LEIBNIZ), but it is not really our concern: World and Time we live in are for us uniquely the world and the time which the God has created. The question, why this world where we live is impregnated with imperfection (especially allowing for the manifestation of anti-values beside the values) has been dealt with at length in English elsewhere[2].

*

1. TIME AND WORLD

The success, without precedent, of Newtonian Mechanics is at the origin of the idea of a time which elapses from $-\infty$ to $+\infty$ always identically the same. Such an idea was not current either during Antiquity (especially among Greek thinkers), or during the Middle Ages, or among foreign cultures. Nor is it truly in conformity with contemporary physics. Indeed, we know that Antiquity attributed to time a cyclic character. And one of the most outstanding differences between

[2] Cf. A. MERCIER, *The Problem of the Imperfection of A World, Itself Created by a Perfect God*, in Foundations of Physics, Vol. 22, No. 2, 1992, pp. 205-219.

thought of Greek origin and thought of Judæo-Christian origin concerns their diverging ideas of Time. For Judaic thought, Time manifests itself in Holy History, - which is linear, although studded with traps, - the history of a people as elected by God, starting with a beginning: the creation of the world; thus therein, time is comparable with half a straight line which is not, and will never, be completely laid out. For Christian thought (itself, it is true, issued from the Judaic one but differing radically from it in its own way, for instance and notably because it rejects the notion of an elected people), Time is an existential line which stops at a terminal: the *eschaton,* as valid for all the peoples of the Earth, and it does not make any sense to either ask oneself what Time was 'before' the creation of the world, since the beginning of Time and the creation of World coincide as if, - to talk like the French philosopher R. POIRIER, God had approved of that coincidence; or what it will be after the *eschaton,* since then God's eternity will prevail; so much so that, for Christian thought, Time looks like a segment of a straight line, yet made of two parts: one before the coming of Christ, the other after that 'zero', unique as it is and of the utmost importance: for only this second part is laden with the whole value which temporality can at all assume, namely that of the *messianic* times.

Let us not forget, that in those days the zero was unknown (it came later to the consciousness of the West from India), and that one has falsely counted, by dating this *parousia* (itself different from the final *Parousia* due to take place on the day of the Last Judgment), at the end (not the beginning) of the Year One. Later, Islam, - arising firstly as a heterodoxy of Christianity and of Judaism (there are even scholars who consider it as the first attempt towards a Reformation of Christianity), - yet very soon turned into an autonomous religion, would shift that zero to the hegira, for it does not acknowledge the title as Christ given to JESUS while accepting the idea of a resurrection and of a last judgment. Still later, the atheistic ideology of the French Revolution, laden as it was with a malignant and arrogant joy, would substitute for it the Year One of the Republic, thus repeating the mathematical error consisting in ignoring the zero, and foretelling - not messianic times, but - purely human times of welfare for the future of peoples and the entire human society[3], as if this society were 'the most important' in World.

Moreover, from Christianity onwards, not only the creation of the world and the beginning of time coincide, but Time and World are One, - a belief (if not even a *thought*) which is eminently paradoxical as considered by a mathematician of strict, yet narrow-minded observance to whom 'one plus one equals two'. And yet, that thought holds not only in metaphysics, but even - all precautions being taken - in physics, as we shall see presently. In other words, temporality and mundaneness are synonymous, - and they will remain

[3] One of the ideas of the Revolution - which by the way has been taken over by the ideologues of the Marxist-Leninistic revolution from the very events in Russia in October 1917 - is that the revolution is an *efficient* and primary cause of a process (from then on a *historical* one - and that is a secularisation of the Judaic idea) recording the irreversible process towards a greater welfare of Society. According to ARISTOTLE, there were four causes: formal, material, efficient and final cause (within the scholastic classification). In our days, formal cause has disappeared; it has been replaced by logical anteriority. The idea of a material cause has been transformed into universal materialism within the said ideology which has, on its own, put aside the idea of a final cause and kept only that of the efficient cause, the paradigm of which would be the revolution; hence the distinction between time *before* the Revolution and times *after* it. Contrarily to Christianity, that break is not marked by the parousia of an exceptional Being (inspite of names like MARX or LENINE), but by the socio-human decision to entrust the people(s) with their own leadership. This means that *from then on* times are revolutionary, whereas for the Christians, times are, since the coming of Christ, 'messianic'. In addition to the difference between the sacred and the profane, that substitution amounts to installing a (revolutionary) movement indefinitely prolonged and constantly directed from the present towards the future (in which - not the eschaton but - the dictatorship of the proletariate is being installed), in stead and place of a (messianic) steady motion of reception of the divine inspiration as oriented from the to-come towards the present. In both cases, the past is of no great significance. The messianic case is compatible with an authentic mystique, the revolutionary one isn't but with a secularized pseudo-mystique. (With regard to the significance to be given, in a new comprehension of what is meant by 'the absconding god', see also the end of the chapter.)

so as late as in our contemporary language, - in that they characterize the state at one and the same time miserable and full of hope of the human creature, however not without leaving open the possibility to distinguish the two terms of that identity: That World has lost for Man its original destination as paradise, is due to the fact that it is the source of 'temptations', yet since the new times are messianic, the hope for paradise does prevail in it.[4]

In order to carry out the separation between that Time and that World, one must yield to an adequate practice of efficiency. In the course of centuries and millennia, man has invented and developed mainly three such practices according to the domains of their respective application, even though he had not an explicit consciousness of this task to complete such separation; they have in common the fact that they constitute the means at the disposal of man to overtake the mastery of what he believes to be the future or the to-come. They are: within religion: prayer, kind of a technique of invocation of the divinity, the Occidental paradigm of which is the Lord's prayer (with the typical use it makes of the adverb *as*), the *'auṁ'* ... being its Oriental paradigm about which one can read in the *Mãndukya Upanishad* (I)[5] that it is "the past, the present and the to-come ...", - then within metaphysics: dialectics, already preformed in the *Veda,* practised by PLATO, perfected by PLOTINUS and developed more and more until our days[6], - in physics: mathematics, which will by and by invade all sciences and give the - albeit illusory - impression that they will supersede them (a

[4] The Revolution claimed to sweep away these precepts, however not by negating the identity of Time and World, - for, in its ideology, the sequence of times and the evolution of World will become one within *historical materialism*, - but by substituting for the messianic hopes the revolutionary assertions of an indubitable efficiency of the doctrin of *dialectical* materialism.

[5] Cf. J. CH. WADIYAR, *Dattatreya,* (London, Allen & Unwyn, 1957, p. 112ss.)

[6] Cf. A. MERCIER, *Metaphysik, eine Wissenschaft sui generis,* (Berlin 1980).The revolutionary ideology makes a steady use of dialectics, falling however into the error to believe, that instead of its being an implement, it be the natural functioning of the very material reality.

little like dialectics tending to supplant metaphysics, or like prayer supplanting religion).

2. A World of Things, or a World of Love?

Indeed, that identity of World and Time has been transposed into the whole of modern science, as is best seen in the example of physics. Even though one would abstract the human creature from the unity which they constitute, Time and World remain one[7]. It is not the case, that this transposition be the conscious deed of the founders of Science; yet if one compares the Christian attitude towards World with the Greek attitude which culminates in PLOTINUS, one does not find in the Greek attitude any hint as to the origin of that identification, since one only encounters the aversion which abhorred the fall of the soul into the subsoil of the *κακὰ κακῶν*. For the Christian, the world of matter - even though it coincides with what was left over after the loss of Paradise - is holden for full of promise since it is the work of God who put Man in it to live there and within love from the Redeemer's very first parousia on. However, the big difference between that world of *love* and the world of physical and natural deeds dwells in the prevailing incommensurability between those two attributes: love is everlasting, or even better it is self-creating and capable of producing miracles out of nothingness[8], whereas any physical action is submitted to attrition, itself of a double nature: the impossibility to make anything with nothing, and the ineluctable dissipative friction within material arrays, be they natural or artificial. Greek thought had never been aware of these differences.

[7] They are one even in a higher degree, because as long as Man is part of it, he inbreathes into it his ontological intuition which has him catch a glimpse of something else than just the plain temporality of the world, namely the transcendent.

[8] As if love were capable of drawing, in order to put them into the world, things that are not (yet) there by fetching them in the to-come, or shall we rather say: capable of letting become evident in time such and such *eventus: τέλος* rather than *συντυχία*, as if love (*ἔρως* to SOCRATES) found admittance at the To-come's master's (*μοῖρα*, even more than ZEUS himself).

So it was necessary to wait until GALILEO, himself a very faithful Christian[9], yet endowed with superior intelligence, to prescind from all religious concerns and descend from the metaphysical down to the physical plane in order to get free from that incommensurability and to allow for an *adequate* science to be elaborated (in the Thomistic sense derived from the *adæquatio intellectus et rei*) as the natural analogue of the supernatural unity of Time and World.

From then on, and since its very beginning, the said science has prevailed as a doctrine according to which time and matter - since World is nothing but matter - do not constitute but one and the same thing, for the simple reason that if matter were not somehow submitted to a change 'lasting in the course of time' in order to be different from moment to moment, it would be impossible to notice that it *is*, and that if matter were not by nature susceptible of change, time could in no way get into the awareness of any consciousness. Such conception is evidently much more advanced and profound than ARISTOTLE's, according to which time is the measure of motion. This mutual implication of the temporal and the material reflects itself in a *conceptual* implication 'Time-and-Matter' (but *not* 'Time-and-Space') which called for a transcription into a scientific explication *of* time and *of* matter allowing to distinguish them, yet without rendering them totally alien from each other, and this required the application of an adapted mental implement which has been found to consist in infinitesimal analysis, also called 'calculus', as the 'modern' phase in the development of geometry (inaugurated by NEWTON and LEIBNIZ), and finally to consist in the whole of mathematics.

---

[9] Consult his correspondance with his daughter Sister in religion in MINA VALLETTE: *Sœur Marie-Céleste, fille de Galilée,* (Ed. de la Baconnière, Neuchâtel 1942).

3. From Ontology to Aionology

That the crucial problem of the conceptual separation between Time and World did not arise for the Greek is due firstly to the fact that it could not arise within the whole of their thought, since it does not seem to have occurred even to HERACLITUS, however serious his attempt may have been to make out of philosophy an *aionology* and not a pure *ontology* as has been the case for all other Greek thinkers[10]. Moreover, even though the Greek may - without being conscious of it - have possessed the necessary implement in its primitive form, that implement was not yet enough elaborated to suffice for the task, since it was available under the form of synthetic geometry only as completed by EUCLID. The Greek had neither algebra, nor above all analysis at their disposal, and their arithmetic was slight, just sufficing to the establishment of PYTHAGORAS' acoustics. That explains why, once Antiquity being at its end, the problem of distinguishing the material from the temporal arose in one camp, namely for the new born Christian thought, whereas the elementary means to solve it were in the hands of the other camp, viz. Greek thought as on its way to ... say, if not to collapse, yet to congeal into that sort of final apotheosis residing in Plotinism.

Modern thought is said to begin with GALILEO, DESCARTES and others. Now, GALILEO has read PHILOPONOS, and PHILOPONOS was, although in his own way, a Christian, whom his monophysitism compelled to fly from Alexandria and got him put on the Index, so he went eastwards and wrote his late works in Syriac, which explains that his thought would be integrated into later efforts of the so-called Arabs, so that when it is said that the Arabs have helped towards the development of a modern scientific thought, one should truly

[10] Except perhaps SOCRATES who in that respect did not either make it an aionology, but an anthropology.

refer to PHILOPONOS who not only wrote *Contra Aristotelem* but also had the premonition of external actions having an influence upon that which NEWTON was to call, much later, the *momentum* of bodies in motion. Of course, at that early epoch of the beginning of the 6th century of our era, and as late as the eclosion of modern science and even later - say up to D'ALEMBERT as a precursor of LAGRANGE - the only geometry available was EUCLID's which, by stating its axioms, defines a space called Euclidean, being metrical, three-dimensional, the coordinates of which are real numbers and which is free from any curvature. For the ancient scholars, that geometry was of a synthetic nature. The analytic translation which DESCARTES made of it kept, by reenforcing it, its metrical and Euclidean character measured as it is with the aid of real numbers; but in the meantime, two or three mathematical concepts, beyond the zero, had come to the knowledge of scholars: negative, imaginary and complex numbers, the continuity of natural real numbers from $-\infty$ to $+\infty$, and algebra had been definitively elaborated by VIÈTE, one of GALILEO's contemporaries. It was all that, which allowed NEWTON to break with the two traditions: the Greek and the Christian ones, while at the same time actualizing their synthesis in reference to the problematics of the foundation of a fit physical theory, since he in his conceptualization gave up the idea of postulating a beginning: *Creation*, and an end: the *eschaton*, the end of a physical time which he postulated to be real and continuous, which he declared to be 'the' independent variable in the sense given to such by mathematical analysis, and of which all the physical properties of bodies are functions, determinable as they are by integration within the frame of infinitesimal analysis which he invented expressly to that purpose, and that he conceived of geometric space as used by that analysis as the *sensorium Dei*. However, the geometry he used was still Euclidean geometry, the only one known in his time.

It is well known that later on, LAPLACE was to make the inference from the thus founded *mechanics* to a determinism which bears his name. NEWTON himself was not a determinist in LAPLACE's way, on the one hand since he did not think that mechanics as he had founded it was fit for universal application to everything going on in the world, on the other hand because he thought that man remains free to place himself on a spot of his own choice in the said Euclidean space in order to throw a projectile with a chosen velocity in a way such that only the future of that projectile was thereupon determined. Newtonian mechanics is a doctrine of the future within the limits where it is a theory of ballistic predictions. The idea of initial data separates a past from a future, and that past of a mechanical nature is as such void of the history of the said projectile, though the latter evidently had to be fetched somewhere: a stone in a quarry, a shell in a factory, an apple on a tree. As the Newtonian mechanical future, one has to understand the collection of possible numerical values which follow the zero of time of so-called initial conditions as freely chosen in a way conforming to the conditions for the preparation of a scientific experience. One can of course ascribe to God the demiurgic care to have thrown once for all the bodies which constitute the material world where we are; this amounts to a model of Creation which, however, says nothing about the spot whence God could have fetched these bodies.

4. *Sensorium Dei,* OR *Sensorium Hominis*?

KANT believed he was right to declare Time and Space to be '*a priori* forms of our sensible intuition', thus degrading Newtonian space from *sensorium Dei* to *sensorium hominis* (as an exterior form of intuition), and by elevating Newtonian time from the status of independent variable to the rank of human sensorium too (as an internal form of intuition). In doing this, KANT meant to establish the apodictic truth of Newtonian mechanics as if there could not be another true mechanics. Luckily for him, the *Critique of Pure Reason* was published few years (only) ahead of the publication of LAGRANGE's *Mécanique analytique.* For indeed, LAGRANGE's mechanics has shown that Euclidean three-dimensional space is not the best frame to have recourse to in developing not only mechanics but the whole of physics, and just one generation later HAMILTON was to show that it even is preferable to refer to a space called phase space, the number of dimensions of which - necessarily even - oversteps in any case 3 and depends on the system one is to describe. That space would later in the 20th Century be used by NIELS BOHR as a paradigm in his elaboration of the elementary quantum theory and was in its turn replaced by a reference to HILBERT-space with its infinite number of dimensions in the theory of quantized fields. All this implies that the apodicticity as put forward by KANT collapses and that so-called *a priori* forms lose their meaning.

Moreover, beside this analytic evolution in the conception of matter and time, one knows that the 19th century has experienced the successive introduction of various geometries which have deprived Euclidean three-dimensional space of its assumed physical priority, so that today one cannot maintain that there is - or especially 'exists' - such a space that would 'contain the world'; for indeed, any space is never *there* 'before all other things' in order to contain them; its sole priority is its abstract and axiomatic construction which however bestows upon it nothing but the statute of being a construct of the mind. Of course, nothing prevents God to be Great Geometer; on the contrary: ἀεί ὁ θεός γεωμετρεῖ, as PLATO said. Time, on the other hand, has seen its rôle as independent variable growing with enlarged importance and has completely lost the status of an *a priori* form of sensible intuition.

5. Four-Time and Supertime

Yet when we simply say 'variable', using the word in the singular, we do not take enough precautions, for two reasons. The first reason is, that the intrinsic structure implied by that variable 'time' depends upon the physical theory by which it is used; the second is that it is not necessarily one-dimensional and thereby comparable with the simple succession of the real numbers of which already **Aristotle** had a foreboding.

This second reason is illustrated by **Einstein**'s theory of relativity. We know that, in contradistinction to the Newtonian perspective, where all mechanical properties of bodies were functions of the one linear Newtonian time, physical magnitudes are in the Einsteinian perspective functions of four variables, themselves analogically assimilated to the three old orthogonal Cartesian coordinates of a Newtonian space and to the old Newtonian time itself of which it is often said that it constitutes the fourth coordinate of *space-time* under the condition that it be converted into a pseudo-length by the means of a factor equal to the velocity of light multiplied by $\sqrt{-1} \equiv i$, a velocity itself assumed to be invariant against new transformations called Lorentz-transformations. That circumstance implies that Einsteinian physics in principle never treats of magnitudes other than physical fields (but not properties), and that its most general problem differs from that of Newtonian physics, for, instead of attempting to determine mechanical properties of systems as functions of Newtonian time along trajectories of their constituents by integrating ordinary differential equations and taking initial conditions into account, it looks for a description of fields of all kinds as functions of the complex of the four independent variables, which means that this four-dimensional complex assumes there the epistemic rôle which linear one-dimensional time assumed in the Newtonian theory. Now, these four variables constitute, or better define, *space-time;* consequently, space-time is not a generalization of space, it is a generalization of time. 'Einsteinian time' is a four-dimensional complex which can be called four-time or *quadritime*, the structure of which is defined by the axiomatics of the light-cone. These conclusions are valid for special relativity and for general relativity as well, even more so for the latter where what is called proper time loses its immediate significance. (Proper time is not the independent variable in EINSTEIN's theory, it is a convenient construct.) Moreover, one has later proposed still more general theories, for instance in order to unify gravitation and electro-magnetism, which make use of supertimes possessing more than four dimensions and whose axiomatics go beyond that of the light-cone.

Consequently, the time of physics is not necessarily analogous to the plain and continuous sequence of numbers with their axiomatics; its intrinsic structure depends upon the physical theory one adopts, and this illustrates the first reason just involved a moment ago. One can put it like this, saying that each well elaborated physical theory is an original conceptualization of time or rather of the mutual implication between time and matter, and one could, by generalization, go as far as to postulate, that any well elaborated scientific theory, including the domains of biology, of psychology and so on, ought to assert itself under that same form of a coherent conceptualization, - for psychology for instance, it might be a conceptualization of the mutual implication of the temporal, the corporeal, and the mental, - yet in a case limited to its own field of application, since the materio-temporal implication seems to be indeed universal and consequently to serve as stage for the scientific development.

6. The To-be and the To-come

Bergson has pleaded with a well-known vehemence against any such mathematization of time - of a time, he says, of which we possess the *intuition* and which is *duration*, in French *la durée*. It is not known whether, before he died, he intended to forbid a new edition of his book entitled *Durée et simultanéité* which had been published in 1922, and where - with a rare circumspection indeed - he had revolted against the Einsteinian conception[11]. As far as we are concerned, we must understand such a revolt, - not simply against a particular conception like the Einsteinian one, - but also against *any conceptualization* of time as such, because it automatically makes out of it less, and consequently something else, than what time acually is, just as, by the way, the whole (never to be completed and achieved) of physics will never be the physical world in its total reality[12]. For, there is nothing more general than, or at least as general as, Time, except the To-Be. Heidegger has explained this very clearly in the booklet which reproduces a lecture entitled *Zeit und Sein*[13]. This meets indeed very nicely our presentation of time as independent variable in the sense of mathematical analysis, even though it is 'still more general', for, one would also say, in mathematics: *t* be the time ..., without asking[14] *who* gives it as an independent variable. One must, - says Heidegger, - be satisfied to declare: *"Es gibt die Zeit"*.

Such a bringing together between Time and the To-Be gives us the opportunity to make a remark concerning existence. A difference between Greek and Christian thought seems to

[11] Cf. Henri Gouhier, *Avant-propos*, in H. Bergson, *Mélanges*, P.U.F., Paris 1972.

[12] R. Poirier has insisted upon that difference between the construction of a *second* reality and the *prime* Reality of World (cf. his contribution in *Philosophes Critiques d'eux-mêmes*, vol. 13, Peter Lang ed., Berne-Fkft/M.-N.Y.-Paris)

[13] Published fourty-two years later than the famous *Sein und Zeit.*, in English *Time and the To-Be*, a booklet where he explains, that the To-Be and Time are *given*.

[14] To Heidegger, such asking is meaningless.

reside in the fact that for the latter, existence is not a mere emergence of the to-be from its essence; rather, it is its coming into, and presence within, time: *παρουσία* (parousia) as well as *οὐσία* (ousia). We shall not here open a long digression about that. By the way, we have behind us S. AUGUSTINE, the father of existential philosophy, as well as the scientific Occidental thought in front of us. The latter, especially modern physics, could not have developed without the inspiration that came to it from the religious, specifically Judæo-Christian conceptualization of time[15], even though NEWTON had to break with it by substituting to its hieratic character a mathematical character. From there came that implacable Laplacian determinism which seems, but only seems to settle not only the future but even the past of beings, i.e. their whole existence, from $-\infty$ till $+\infty$ in time.

And yet physicists, who constantly remake physics by adjusting it along the needs of the problematics considered fashionable, are not that strict about theorems, because physics is not mathematics. Indeed, since NEWTON, one has always proceeded as if God in His capacity as the Demiurge had chosen to let the celestial bodies start from an initial position with an initial velocity or to let a primitive atom explode, which means that inspite of a mathematical contradiction, one conceives all the same of a beginning or an origin of time coinciding with the creation of the world in a past long ago. On the other hand, physicists have never speculated similarly about the end of times, if not in thermodynamics around the increase of entropy which, however, leaves the moment of an end in a total imprecision making it rather asymptotic and not eschatological. Moreover, thermodynamics considers a determinism of its own which differs from Laplacian determinism, since according to it, all

[15] This statement has been uttered for the first time by SVANTE ARRHENIUS in *The Life of the Universe*, (Engl. transl.), London and N.Y., Harper and Bro. 1909.

will end in universal confusion. In other words, if physics claims to be the previsional study of the future of systems, it is found to be the study of what has been the case as well as of what will be the case. And Science is due to keep to the greatest circumspection as soon as one requires it to play the rôle of a 'futurology'.

7. Inconsistency of Determinism

Einstein has considered it useful to establish a tie between his physics which, strictly speaking, has only to do with fields, and the preceeding physics which dealt with particles and other bodies on trajectories: this led him to give, of these old bodies, an image known under the name of their world-lines. A world-line is not a trajectory which could be cinematographed. What is correct is to say that, within Einsteinian cosmology, any body is nothing but a world-line as laid out in space-time. Now, space-time is time: four-time. Consequently that line is such as it is there and not otherwise, and it is impossible to shift and displace it. That impossibility amounts to a radical determinism infinitely more serious than Laplace's determinism. In such a retrograde interpretation of the Einsteinian conceptualization, there isn't even the past, or the present, or the future; there are space-time and the world-lines, but nothing else: taken literally, it makes it impossible to find oneself outside space-time, be it just to write and talk about it, since human beings are world-lines as well and cannot but keep silent (or if they do write and talk, that is part of their determined career including what they write and say). And even though the strict representation of fields be re-established, the same conclusion applies; there is nothing but space-time and the fields, and this is at one go Four-time *and* World, so much so that in a hyper-Laplacian spirit plants, animals, men are integrated into it like stones and heavenly bodies, and their totality is a *pan-cosmos* as undiscernable from a *pan-theon* where all the gods are but one, incapable to determine himself.

The plurality of determinisms shows that none of them can be ontologically interpreted. Determinisms belong to the hermeneutico-epistemic domain. Reality itself is not determined; yet theories as put forward suggest that it is determined one or the other way. The Twentieth Century has seen the birth of quantum theory, of which it has been, - wrongly, - said that it is indeterministic; actually, it allows one to make predictions which are very well verified within the frame of a more flexible determinism than within the other theories. Controversies between EINSTEIN and BOHR on this are well-known. Actually, BOHR was always right, and EINSTEIN obstinate; this is due to the fact that within EINSTEIN's theory, determinism is radical, within quantum theory, it is not radical at all.

8. THE INTUITION OF DURATION

Yet let us come to the will-be intuition of a time as *duration*. Is it not so, that the human spirit likes to refer to it as if to a flow, the flux of which remains equal to itself? Notice that this last sentence does not mean much, for it is impossible to compare together two instants which are inserted at different epochs of a time even meant as intuitive, and because the 'velocity' of that flux: dt/dt, is always equal to *one* so that, on the whole, World, identical with Time, cannot constitute its own clock[16].

However, it remains that one would like to deal with such a flux, and this is so true, that this question is put to every theory assumed to conceptualize time: does such a theory admit an intuition as just envisaged? The answer is: Yes,

[16] The idea that the world resembles a clock is consequently wrong, (*unless* it is a clock as used by God).

within the limits in which we can fabricate a clock based on the principle of each physical interaction as accounted for by each of these theories, and then the time indicated by each clock (do not say 'measured' by each clock, for measurements are made 'in the course of time', and thus time is never 'measured') - the time thus indicated is an image of a capital Time just as a map on two hemispheres is an image of the Earth, or just as e.g. the picture of the *Creation of Man* by MICHEL-ANGELO contains an image of God. This question is particularly important for theories within Einsteinian special and generalized relativity, for neither four-time, nor any super-time seems, to begin with, to allow for it. However, they allow for it as a consequence of an artificial construction which is always possible and is known under the name of 'cosmic time' which, starting from the proposed cosmological model, is calculated under the form of a linear variable interpreted as time of that model of the world. This means that one still is in a position, directly or by using an artifex, to speak of a past and a future in reference to all known physical theories, and precisely in reference to cosmic time does one talk e.g. about a Primitive Atom or refers to a Big Bang. As regards scientific theories in other domains - biology etc. - this question remains open. But let us recall here that, to animals, the heart amounts to a natural clock.

Now, one of the most general theorems having been demonstrated in theoretical physics, is that for all these theories, their implications remain unchanged if the sign of time flow is reversed under the condition that also certain symmetry operations are carried out. That is of so general validity, that it is nearly impossible to say whether the world we live in is the one we begin to describe or the one symmetric to it. It is called the TPC-theorem. Then, taking such precautions, if we agreed to change the sign, i.e., the direction of flow of time, we should not be in disagreement with physics.

Yet on the other hand, we could not any more proceed to the same epistemic interpretations, especially in regard of determinism and, as we shall presently see, of the rather exciting ensuing metaphysical consequences.

9. The Denomination of Time as the To-Come

Several languages of the world, - as Professor G. Redard, former Head of the Linguistic Department of Berne University, has explained, - do not offer an equivalent for the word 'the to-come' constituted by the preposition *to* and the verb *come,* used as a substantive. This might be, in certain cases at least, due to the fact that, for certain peoples, this very idea of a future either did not arise, or constitutes a taboo because it amounts to tempting the Divinity if one were e.g. to ask what will happen to-morrow. However, among other cultural groups, the *future* puts a problem either of conceptual, or of hermeneutical or even other nature which it is reasonable to tackle. Hence, the question arises for us to know not only where the need comes from, to endow the future with a significance, but also what the meaning of the very idea could be within a given culture like the Occidental one which is both of Christian and of Scientific origin.

There was of course, to begin with, the wish within certain ancient peoples to predict; and, among them, procedures were known to achieve such in horoscopes and manipulations as have been practised by haruspices and other soothsayers. Their concern was to *pre*-dict what was *to* come: a transformation of an unknown future into an anticipated present as figured in the concrete present of animal entrails or of other arrangements of things. Yet precisely, the meaning of such transformation is not the one implied by the very idea of a to-come: For, the problem is not, to know in what sense the future is present in the concrete, but in what sense the present is 'to come', that is: that which comes, and which comes 'to meet us'. This latter concept is already better formulated in prophecy, if prophecy be not conceived as a figuration of the future in the present but, inversely, as the projection of the present encountering the future: for instance, for the Jewish prophets, in the descendant of DAVID'S himself present in the Holy History, who would be the future Messiah of the same History. Hence JESUS, for the Christians. Henceforth, the 'messianic times' which will follow and appear as *final* cause of our salvation, but are actually its *efficient* cause because they ceaselessly come and encounter us. Yet, one should avoid the use of the verb 'to follow' because it is precisely still based on the idea of a duration of time - not any more, 'tis true, from the past - but much rather from the present (in the sense of the coming, of the parousia) towards the *eschaton;* whereas the idea about which we should agree is, that time flows from the future (and hence also from the *eschaton*) towards the present.

In other words, what we are looking for are the profound semantic contents of the phrase *the to-come*[17]. Most of the Indo-European languages use a similar phrase consisting of a preposition like *to* (in French: *à*, in Spanish etc. *por*, in German *zu* and corresponding ones for Nordic and other tongues), together with a verb expressing motion, like *come,* in French and other languages derived from Latin: *venir* and so on, yielding the words *avenir* in French, *avvenire* in Italian, *porvenir* in Spanish, *porvir* in Portuguese, then *Zukunft* in German (where *kunft* comes from *kommen*), *tillkommanden* besides *framtiden* in Swedish, whereas Dano-Norwegian uses the word *tilkommende* practically only to designate the future husband, for the future is in that common tongue called *fremtiden*, i.e., the time (*tiden*) which is located ahead (*frem*) and suggests progression. English uses currently *the future*

[17] The question, whether conceiving of Time as the To-come has a repercussion on the Logic of propositions within Language, shall not be tackled here.

as borrowed from Latin, but the Oxford Dictionary accepts *the to-come*, as tainted perhaps with a touch of poetry. Latin does not seem to have used the same kind of locution; it used *futurum*, which is a participle of the verb *esse*, hence a form of the to-be, which sounds 'ontological' whereas the Latin words for the past: *præteritum* and the present: *præsens (tempus)* are built with the help of a preposition: *præter* which already admits the past to which is joined a verb expressing the come or the leave in *itum* from *ire*, or *præ* which expresses the facing to which is adjoined *ens* as expressing the fact to be (i.e. again the ontological position). But, after all, *futurum* means exactly 'due to be', so that the difference between Latin and modern tongues consists, instead of the to-be which refers to an ontological situation, in that the modern languages say to-come which refers to an aionological situation, as if substituting an ontologizing, (Greco-)Latin thought with an aionologizing, Christian thought had happened, a substitution needing its image in language.

However, this explanation would need more confirmation. For indeed, Greek used ἐπιόν, which is the neuter participle of ἔπειμι meaning precisely to come; consequently, taking into account the sense of ἐπί, one encounters in it the 'come towards', the 'encountering', or eventually the happening. One could also refer to the Greek ἐπι-βαίνω which not only suggests the happening, but also the bringing, a letting happen if you please. Now βαίνω corresponds to the sanskrit *gām* which is etymologically connected with: *at koma* in Old-Norse, *kuman* in Old-Germanic, *kommen* in modern German, *(to) come* in English of course. Yet on the other hand Greek also uses τὸ μέλλων which implies intention or even duty, as in μελλό – γαμος: the future spouse (exactly like in Danish *den tilkommende*).

Hebrew does not possess a word to translate the vocable of the future in a univocal manner. It uses circumlocutions.

Yet the *Old Testament* uses certain expressions in order to suggest the times that are still hidden. According to Professor KLOPFENSTEIN of the Faculty of Theology at Berne University, the expression which might come nearest the one we are considering would be *ha'olam habbā'*, which is found again in the Biblic Greek locution αἰὼν ὄυτος for times which are not yet (say: 'those times'), but which however will come. Arabic however, in its turn, has a sense of the future, and uses even, in order to express it (as communicated by Professor C. BÜRGEL from Berne University), the vocable ***mustaqbal***, which can be said to mean 'that of which one expects that it will happen (to us)'. This, by the way, is not astonishing, since the idea of a Last Judgment is not alien to the ***Quran***, even though JESUS is not any more considered in it as the Messiah. Modern Persian, when talking about the future, uses *ayande*, which means 'that which comes', and Turkish says *gelecek*, a future participle meaning 'that which will come'.

From all these ways of expressing the to-come, one gets a common impression of the idea of a time that comes towards us, even as if it were its very intention or at all events its duty imposed upon it by a superior instance which cannot reside elsewhere than at the spot of its source. With all reverence due to BERGSON, this takes us away from pure duration on two grounds, first because duration does not comprise the intentional connotation, second because if one attributes to it the latter all the same, then the orientation which appears to follow is opposed to that which the verb to endure might imply when used in the sentence: Time hangs on my hands, in French: *Le temps me dure.* On the contrary, if we consider the future: the *to-come*, as the source of time, time does not hang on our hands, since it then becomes the bearer of both the hopes, and the threats, of what might happen to us.

It is our belief that this was the purport of thinking Time in the whole depth of its meaning as was the case of the two great Christian philosophers and theologians: S. AUGUSTINE and PASCAL. For them, as well as for the authentic Christian, everything reaches us from the future, and that which thus reaches us, is World as well as Time, in its material and spiritual appearances. For, the identity of World and Time is in no way destroyed in that new perspective; on the contrary, it is strengthened: World is in perpetual regeneration and gets filled with surprises which, from the very present, proceed to the state of its past determinations which from then on can be told in history books. Yet, this also signifies as well that the creation of the world 'does not go back to the most remote past', but that it consists in the transit of Time (and of 'World') to come into the present (the present world) which immediately becomes past, and that it therefore never gets achieved (unless a radical and final *eschaton* stops everything).

But then, that which has passed is past, kind of self-evidence which, all the same, is not unimportant, for what has passed is thus terminated (i.e. has reached its term), and that alone is what is *then* de-termined, and eventually told fragmentarily in books on our shelves. Moreover, what in former days was called the beginning (of World and Time), in short Creation in the old, now obsolete, sense, is now a term, of course the most remote in the accumulation of what is past. And thus in fact does the past not even belong to time itself as a flow. It is nothing but a dustbin (archeologists speak of *køkkenmødding* from the Danish word for the heap of filth thrown away from the kitchen), increasing in size, where the pre-historians can dig. Therefore that term is in any case not an *end* in its sense of a τέλος. The whole problem of teleology has to be put differently: one must drop thinking along the Aristotelean causes; in particular, the old final cause puts on a completely new and *efficient* coat, since its origin is on the side of the to-come of time in regard of the present of the parousia of the *οὐσίαι*, so to speak 'on the right side', while the old efficient cause becomes dubious, for it is not clear how the past in its achieved termination could still - if we quote from LALANDE's famous *Vocabulaire de la Philosophie* - be the bearer "of a phenomenon which produces another phenomenon" or be "the being which produces an action".

Contrary to a steadily more remote past, the to-come is steadily nearer; it is even at any moment infinitely near, even though the present remains distinguished as if by a partition having only a one-way transparency. Such proximity or nearness is being manifested by the unbroken appearance: the *parousia* of all that is given and happening to us[18], part of which is made of surprise, and even of miracle, the other part of which does not surprise us because it either reminds identically of the past like the sun-rise each morning, or similarly as does the renewal of the seasons or the birth of youngsters, or analogically like wars, ups and downs of cultures, cataclysms, historical deeds ... Yet a surprise is a surprise only if it has no analogue, no likeness nor anything identical with it in the past.

10. *Argumentum de Deo abscondito*

Well! Are we to expect the worst, e.g. that the reader will sentence to ridicule the proposal of a conceptual reversal of the flow of time, especially when accompanied by the last commentaries? One cannot help being if not quite taken aback to begin with, yet at least amazed, but also delighted when all this occurs to one's mind for the first time. But then, one not only gets accustomed to it, but also finds great satisfaction in it, because the to-come gets thereby

[18] Cf. HEIDEGGER, *supra*.

an extraordinary obviousness without losing for all that its mystery. After all, all the attempts at approaching the 'Problem of Time' as one finds them spread along the history of the sciences, of literature, of philosophy are never but conceptualizations of Time, accompanied or not by a critique *à la manière de* BERGSON *ou de* HEIDEGGER. We could go so far as to say that from the point of view where we have arrived, there is but one more step to take, - a hypothetical one like all the rest, for sure, - in order to reach the summit of these considerations, namely by personifying the source of that time 'to-come'. But why, - some will ask again, - take that step: why add one more abstruse proposition to all that? Answer: first because one shall, philosophically, never abstain when it becomes clear that this step is the last walk towards a kind of Absolute, and second because if one then would stop where we have arrived so far, the said source would stand devoid of justification, and that would leave us in a kind of ontologico-noetic vacuum. We would then be left with just the possibility to say, in the way of antique India, that there is a reservoir, infinitely full, out of which time flows, like a huge glacier which in melting gives birth to a river without ever shrinking to nothing.

Even though we should be satisfied with the mere supposition about such a reservoir as the instrument of a supreme divinity (BRAHMĀ, ... VISHNU), it would evidently be absurd to consider it - with the intention to agree with the old conceptualization of a time flowing from the past towards the future - as situated before time in the past, for indeed, time having started to run, the reservoir could not be any more before the beginning of time and would have to collapse as a mere contradiction. Consequently, whatever reservoir it may be, it can only remain such by being on the side of the future, remaining thus the source of what comes and constitutes both time and world; then one would be in possession of a somewhat *coherent aionology...*

We could leave it at that; however, the isolated idea of a reservoir as full of time in reserve is really most anodyne (in the sense of the Greek *ανώδυνος*: free of the pains of childbirth), beside the fact that one cannot help asking the question: why that reservoir rather than nothing? There is no reason why there should be one reservoir but nothing to justify it. On the contrary, if the source is made a Capital Person, that question collapses, because it is not any more a question that we can ask, since clearly the said Person is superior to the human person. Consequently the responsibility of that Person's acts is the Person's own and is not the business of human beings for whom at most the 'Leibnizian' problem arises of its philosophical justification, yet as displaced from ontology towards aionology, that is: the problem of a 'prosopodicy' of an aionological nature instead of a theodicy of ontological nature, and this is indeed not bad at all, considering that it opens a totally new quasi-infinity of possibilities for philosophic speculation.

With regard to this, we must understand that, in order to assume the nature of a person (*πρόσωπον*), an *ousia* in the Aristotelean sense must be able to posit by thought one and the same thing several times in a succession and recognize it as identical, so much so that it will recognize itself in that succession in order to outspeak "I". Thus a person is one who is able 'to speak in the first person', and who is not reducible to mere properties and relationships. Then if there is at the source of the to-come a Person in that sense, this Person is the One who gives successive birth to all instants of Time which 'at the same time' are the instants of World and constitute the succession of surprises for the human beings who are their recipients.

It indeed seems that this role: of being at the source - or rather perhaps: to be *the* source of the time which comes to us from the to-come in the shape of World, repeating itself ceasclessly without resembling itself, is the pre-eminent Act as continued identically which makes of the said person and source a Person and Creator of World. But, if this mechanism is to be viable, there must be between the Capital Person at the source and the minute persons at the arrival a *persistant mystery* which finds its expression in the fact that that Capital Person at the source *absconds*, for otherwise, that Person would be no one else than one being among the beings of World, one ('big') person (kind of a ' grand'father' with or without a white beard) among human persons, and at most susceptible to abscond for a time behind a column or behind a cloud or even behind a galaxy, in short at a place where to finally be caught sight of and found, so that that Person would neither any more be at the source of the to-come, nor be *the* Capital Person, Master of Time and World. Hence the Capital Person (Source of times and World) does not abscond in any sort of space, which is in any case a mere construct of our mind; much rather, that Person absconds in time or better 'behind time' - yet: where, behind? - well: behind the to-come, so that the to-come invariably remains the to-come, even though it constantly appears under the form of the present and of the world in its parousia: World is always Present time as renewed by the temporal surprises, among which even Miracle is possible.

And then, the idea of the reservoir is not only of no use; it is null, since World is given together with Time as by the Capital Person to whom we cannot help giving the name of the 'absconding God', *deus absconditus:* One who remains hidden. For this is the condition *sine qua non* for being God, otherwise He would not any more be the God from Whom all comes to us and would immediately collapse into a past where all there is to find is that which is finished and has no other importance than that of teaching us to accept the inescapable.

Chapter VIII. The Insurmountable Intricacy of the Problem of Time

Time, in the classical understanding, e.g. - but not necessarily, - specified as Newtonian time, is not the 'contrary' nor the 'opposite' of the *To-come* as we have attempted to characterize the latter, nor is it actually the same as that, which is also called 'time' in (thermodynamical) considerations of the irreversibility of universal decay.

Newtonian, or any other 'physico-existential' time, i.e. as *implicated* by a precise (dynamical) physical theory, is usually and correctly designated *as a variable* in its mathematical sense: call it t; and, since the publication of Newtonian Mechanics in *Princ. math. philos. nat.,* it has been understood as the - unique - independent variable upon which everything that constitutes the world depends. However, no proof of the identity of times as implicated by different such physical theories has been put forward so far by physicists.

If *the to-come* as we have introduced and commented upon at length were to be a comparable variable as, say, designated by the Greek letter τ, one might be tempted to maintain that 'at the *point of coincidence* of Time and the To-come,' - a point which of course 'changes' 'all the time' and can only be *that* which is called 'the present', - dt and -dτ should be equal, however certainly *not* that t and -τ themselves should be equal, since it indeed is not possible to assign any unique zero value to the to-come, a zero which would have to be the *eschaton,* of which we know nothing else or more than its mere possibility, unless we take the (many) biblical sayings about it for 'physically' granted (which is unjustifiable by reason). Hence τ can only be defined as the indefinite integral $\int d\tau$. Especially, no negative values of τ would make any sense. Quite the contrary, we can for t chose a zero, though it be in any case arbitrary, for a 'true' zero would have to coincide with a beginning excluding any former time, hence: e.g. the birth of JESUS may serve as $t = 0$, and then negative values of t are also meaningful, down to numerical values say beyond -10^{10} years. Of course, a 'birth' is pretty vague since, for a mother to be confined can last several hours, and any 'birth' lacks precision. In practice, any chosen zero is endorsed with imprecision, like any other historical date by the way, since no mathematical variable as more primitive than t can be used to fix the zero of t. Such an inevitable imprecision is a challenge to any physical theory which, owing to the choice of its mathematical background, is supposedly built on rigorous exactitude.

Neither Time, nor of course the To-come, are ever measured, owing to the very reason that we cannot dispose of them: we dispose neither of history, which evidently cannot be repeated, nor of the future even though we try by all means to predict it. All we dispose of are *watches*, which should never 'stand still' at one and the same instant and indeed do not stand still as long as they keep going thanks to the potential energy contained in their work, and they stop being working watches as soon as they stand still; therefore they may show time (though not measure it); but they do not even show the to-come.

Strictly speaking, each physical theory implies, or even defines *its own time* as based on a particular genuine interaction, and allows in principle the fabrication of a watch meant precisely to show that particular time.

One kind of time however, namely the To-come, is not mastered by any physical theory. And one physical theory does not allow for the fabrication of a watch and consequently does not rest on a similar time that accompanies, or accounts for, change in the 'order' of things, but on 'a time of its own' which will (or should we say: shall?) lead *to the end of things,* i.e., to the end of their 'existence proper'. That theory is thermodynamics. For indeed, thermodynamics is not the theory of (the workings of) a particular interaction like gravitation in Newtonian mechanics, electromagnetism in Maxwellian electrodynamics , ... Thermodynamics is the theory of supposedly isolated systems, even the theory of the finality of that isolatedness, if you will. In praxis, thermodynamical machines - smithies and similar works, motors, ... and the bodies of living beings - are more or less badly isolated, the best isolations being encountered in freezing devices, e.g. in modern deep-freezers. The only system which is rigorously isolated is the World in its totality. But no watch can be fabricated upon the principle of that isolatedness, and consequently the world as a whole is not a watch, unless it is one to be used only by the Creator of the world, the Demiurge: God if you please, Who alone 'knows' how to handle it. But even such use by God is equivocal, because the To-come is such, that at every instant of its evolving, a 'more' is by continuous creation added to World, so it does not exactly recall the simple succession of the numbers, rather it reminds of something like an exponential function as if at each 'instant' of τ, the increase in τ were a function of the corresponding t.

Strictly speaking, thermodynamics is not of the same kind as other dynamics; it is even not a 'dynamics' proper, for it is not a theory of one or several forces or interactions (Greek: *dynamis,*) but a theory of things in their bulk: a *plethorics* (from $\pi\lambda\eta\theta\acute{\omega}\rho\alpha$: plenty) perhaps. In such respect, it is opposed, if not absolutely, yet factually, to the consideration of the creativeness which accompanies the to-come whose action, on the contrary, is to suggest the novelty of birth within the dynamically ordered change of state of things existing in the world. So much so, that we could establish a threefold correspondence along which: (1) the To-come or 'time of creation' (or should we rather say: of creativeness?) as the *bearer of the birth* of *things* (Latin: *res*, whence: 'reality'); (2) physical time, - or possibly times along various dynamical behaviours, - as current duration(s) serving as the *support(s) of the existence of the said things, i.e., of reality;* and (3) thermodynamical time (or perhaps better 'tychical' time - from the Greek τυχικός: fortuitous, accidental) as the *bringer of death,* do run parallely.

The to-come itself, in contradistinction to the times implicated by dynamical theories, is - 'for the time being' at least - neither implied, nor defined by a specific physical theory to be learnt from a text-book or a published paper. Up to the end of the 20th century, the various times defined by various dynamical physical theories have been considered identical; even that tychical time of thermodynamics has been considered identical with the times of dynamical theories. But the to-come, even with a reversed sign, has not been: it has not even received any consideration on the plane of physical research and theoretical approach. Identifying those times with one another implies in particular, that at a common present, all their differentials are equal (in absolute value). If physics is conceived as the elaboration of an all-embracing theory or complex of theories about particles, it may be that one and the same particle possesses at least mass m, (electric) charge e and possibly further 'properties' like spin and what have you. We do not know whether such properties are independent from the behaviour of these particles in the course of the time involved in their specific behaviour, eventually through the intermediary of some other variables as apt to describe their temporal state, e.g. a velocity if Cartesian-like coordinates are used to designate what is currently conceived as their position with regard to given frames of reference. This very problematics was not even envisaged by the founders of early theories like mechanics and even electromagnetism. Much rather, they tacitly accepted, that there 'runs' a time of universal significance, the same for all things in the world, as suggested by the (assumed) eternal repetition of the apparent motion of celestial bodies, which implied in fact that physics was from the outcome conceived as a cosmology.

It is on purpose that we say 'there runs' a time, and not 'there is', for the to-be should in no case be confused with time (whether 'physical' time, 'tychical' time, or the to-come). The languages which suggest such a confusion are unable to support and distinguish from each other an ontology and an independent aionology. (In this respect, the German "es gibt" or the French "il y a" are much more respectful of the need of both ontology and aionology.)

In former historical times, it was tacitly assumed - because intuitively, yet naively believed as obvious - that there runs indeed one single time: the same for all existing things, especially for all experiencing, thinking and feeling beings, i.e. at least for all human, and possibly other, beings, and that single time was felt as running from what men call the past towards what they call the future, these two segments of time being connected with each other at a 'point of time' devoid of duration called the present, implying a steady increase of the totality of the past and the steady decrease of the totality of the future, even though the beginning of the past might go back to infinitely earlier events and the end of the future shift itself to infinitely later events, the same earlier, or later, events for all beings, there being a universal *temporal* background against which World in its detailed totality would change and evolve.

Now, the above, long and tedious comments might, to the taste of many thinkers, preferably have been left out since they make explicit though they never totally cover the intuition human - and possibly other - beings have of Time. But they also throw a serious doubt upon the rightfulness of the pretentions of such a will-be intuition. There is indeed one peremptory reason for explcitly developing these comments, namely because the contents of such 'intuition' have been rendered dubious by the raise of the Theory of Relativity, in particular with regard to the assumed explicit universality of time. For indeed, while it had been thought originally that time - or at least such and such time interval - is a datum susceptible of measurement and that its measurement was to deliver the same results independently from the performer of this measurement, precisely this was with the coming of Relativity Theory found to be a wrong assumption, for will be measurements of time intervals were then found relative to the subject or observer, more precisely to what is called its motion, which made the traditional Aristotelian definition of time a fake. Only because relative motions of things (and observers) in the world (except photons) are very much slower than the propagation of light can one as an approximation still talk about a quasi-universal time; strictly speaking, the idea of a simultaneity of events does not make sense any more. Even the dating of events has lost its rigour, e.g. the battle of Waterloo would not 'have taken place at the same time' for England and for France if these two countries (as subjects and 'observers' as well as fighters) were or had been in very fast relative motion, i.e. at a relative velocity approaching c. Physicists can compare their sayings and especially their measurements only because they all live on the solid Earth and walk or travel 'slowly' at its surface.

Consequently, none of the times defined by the various current physical dynamical theories is 'universal' and none of them allow to define, or even to conceive of, the simultaneity of two happenings as registered by two distinct things (e.g. two apparatuses measuring each a property of one autonomous physical system).

On the other hand, whether the to-come is endowed with universality, is not an easy question to answer. The To-come is not just time t reversed in its arithmetic sign. The To-come *and* World as a whole are related, and their relationship is 'creation'. Here, like in other quoted cases, the three notions of the to-come, the world, and creation are related in a way such, that, - very schematically, - if C stands for Creation (like $\vec{f}$ for force), W for World in its entirety (whatever is meant thereby!), and τ for the To-come in its universality (whatever again is meant thereby!), a fundamental postulate might hold and apply in the schematical form $C = dW/d\tau$, analogous to $\vec{f} = d\vec{p}/dt$ in Newtonian mechanics. In such a case, C would have to express the intentions of the Demiurge and might be a 'function' of all the parameters which are necessary to describe the state of W at τ, and both W and C might be supercomplex entities very much more elaborate than any vector, tensor or hypercomplex number used in some geometry or algebra. In that sense, and frightfully schematically, the Demiurge is a Great Geometer or Mathematician.

Connected with this, τ would imply a universal simultaneity. Yet, as a matter of 'fact', only the Demiurge could and 'would' conceive of it, as if the above formulation of the postulate were issuing the pregnancy of a world as inseminated with the to-come. Saying that is awfully schematic, and may sound ridiculously anthropomorphic at the same time, as if not just the Earth (in Greek *Gê*), but the (huge) entire World (*Cosmos*) or rather its infinitesimal change dW would stand at the numerator of the right-hand side of the equation. The desperateness, for human beings, of finding the 'true' functional expression of C is notorious! For it would have to subsume all sciences and reach a divine knowledge which simply oversteps the bounds of human cognition.

*

Finally, in order to avoid a misunderstanding which might lead to a criticism, a clear distinction must be made between what has just been said in this very chapter, namely that Time appears to part into *three aspects:* as 'creational', 'physical', and 'tychical' time, and what had been said at the opening of Chapter II where, in the first Section, four possibilities were found available to a notional interpretation of Time, owing first to the two characterizations coming from either the (historical) past (index $_{hi}$) or the future (index $_{fu}$), second to each being attributable to either the proper existence of one single existing being at at time (index $_{pr}$) or the world as the whole of what exists (index $_{wo}$).

For, it might seem at first sight as if a contradiction between a three-partition and a four-partition of one and the same 'time' arose. Yet there is not such a contradiction, because the four possibilities which were discussed in Chap. II and the three aspects put in evidence in this very chapter are obtained along two different approaches of comprehension and consequently of discourse: The first one *sub specie rerum distinctionis,* where Time is the very prime idea of the Given, - the second one, on the other hand, *sub specie actionis exitus,* where Time is the very primitive idea of the Bearer of Poiesis.

In this latter case, one is confronted with the intention of the *Author* of the action leading to one of three possible issues; whereas in the former case, one is led to the search of possible distinctions in the characterization of the temporal behaviour of Reality. The latter: the consideration of the distinctiveness within things, is a typical human approach, which appeals to Man's analytical mindedness towards all that is worldly: it confines to *the physical.* The former: the consideration of the determinativeness within the wealth implied by temporality, cannot but be of divine concern, insofar as God is the source of Time: it leads towards *the metaphysical.*

Therefore we find ourselves led by the introduction of such distinction back to the heart of our considerations: *the problematics of the relationship between God, World, and Time.*

Chapter IX. The Illusions of Discourse

The Chart on p. 69 puts at the level of 'pure form' within the domain of the incommensurable the supposedly comparable terms: logics, poietics, and deontics.

They refer each to a distinct value which has each been designated as cardinal, tacitly implying their independency beside their supreme character. These values are truth, beauty, and the good.

But contemporary logics in its formal exposition tends to reduce discourse to atomic propositions or sentences of which it must be decided, which 'truth-value' they possess, and such truth values have been classified as belonging either to a two-valued logic admitting only truth and untruth, or to a three-valued logic adding a third possibility admitting of, or leaving room for, doubt, indeterminacy or what have you; and then the question arises, if logic is not by force restricted to two values, whether the highest number of its possible values should stop at three, and if so, whether the third possibility is of a univocal linguistic signification, or whether by being neither clear truth nor clear untruth it admits of a discontinuous or continuous set of values between the clearly true and the clearly untrue. To ask whether it would make sense to imagine a one-valued logic would amount, it seems, either to say that any sentence is true, or to say that any sentence is untrue, whatever be its construction in words. If this did make sense, one should - it seems - rather declare that any such sentence is untrue, for even in case it is extremely probable that it be true, this would still be a kind of untruth, at most an approximate truth, - and we remember that, according to BERTRAND RUSSELL: "Although this may seem a paradox, all exact science is dominated by the idea of approximation".

Now when teachers of formal logic are asked to give examples, they practically always furnish seemingly simplest cases like: 'Socrates is a man', which is said to be 'true', whereas 'Socrates is a woman' would be untrue. Yet first of all, using the present tense asserting that Socrates is, or is not, such and such, is subject to a serious critique, since SOCRATES is no more alive and therefore cannot be or not be this or that 'now'. The world changes, and much of change, even all of change in World, consists in things not being any more what they have been.

Secondly, even the will-be 'verifications' operated in science are not, say: 'a true establishment of truth'. They are the establishment of *facts*, and moreover they necessarily are such establishment within the limits of accuracy allowed by the arrangements of the experiment made to perform the said verification; hence no will-be 'universal truth' can be made sure by 'verification'; whereas only the eventual falsification makes sure the falsity of such assumed truth. And this is by the way a main reason for assenting to POPPER's postulate of falsifiability.

Also the 'truth' which a witness or an accused person are expected (and have sworn) to tell at a court of justice is not truth as that cardinal value which is meant in our discussion and implied in the mentioned chart. For if the witness or the accused tell the untruth i.e., a lie, they will be punished for that, for a lie as a lie is opposed to 'the good' much rather than to 'the true', and this regards morals, not science.

Thus at the same time a shift is operated from the domain of truth to the domain of the good as if there were a sort of identity between the true and the good, the image of which is found in the identity between a lie and badness i.e., evil.

And thus final, total, profound Truth is something which has no contrary; or rather no other than evil, just as the opposite of final, profound and total goodness is of course also evil, as well as - to mention the third, middle cardinal value - the opposite of beauty is evil, not just uggliness, even though it could be called horror or some other name.

On the other hand however, World can, by some at least, be found beautiful, or uggly, good or evil, but not easily true or untrue! Yet even these are doubtful assessments. First because World changes and can by the same thinkers be estimated differently from time to time, moreover because no one has at any moment of time a total grasp of World, so that rather the presence, in World, of horror and/or of evil can be asserted in contrast to what is held for beauty or goodness. The reason for these unwonted asymmetries rests on the fact that, whereas truth and untruth necessarily refer to Discourse (especially, but not necessarily, *about* World), beauty and the good refer necessarily to either World itself or parts of it (occasionally to the 'world of discourse' which sometimes, as intended to be poetical, may or may not be found beautiful) or as either praise (ἔπαινος) or calumny (κακολογία) may or may not be found good. Yet in any case the presence or absence, in World, of truth or untruth, does not make sense if not clad in discourse; but unfortunately there is no universal discourse: as the same outspoken by all, not simply because 'all' do not speak the same language (translation could be made to reach a kind of universality of discourse), but because they constitute a temporal, hence changing ensemble whose opinion is known to fluctuate and to replace opinions found obsolete by others held to be preferable. Thus discourse itself fluctuates between agreement and quarrel.

Hence the problem cannot be to say explicitly, what truth, or beauty, or goodness are, for they are the cardinal values

- and we should know already what a value connotes from a postulate as enunciated further above, and consider precisely these cardinal values as the possible ones and look for a reason why there are the three cardinal ones and eventually no further ones.

The fact that sentences as uttered in discourse cannot but be found either to coincide with factual truth or not makes out of discourse a very weak instrument, perhaps even an illusory procedure in the search of a truth which is meant to coincide with a cardinal value. For indeed, if Value is an element of a postulate of thought as meant to connect three such notions (value itself, quality, and quantity; see above), like force is one element in a postulate of knowledge as also meant to connect three such notions (force, momentum and time, ibid.), the possible values of, say, truth, beauty and the good are comparable with, say, possible forces for which laws of force can be proposed, which could make believe that something like a law of truth, a law of beauty, and a law of the good should be the possible objects of discovery of an axiological (instead of a physical) search making it precise, under which conditions either of these can manifest itself, needing parameters to do so. But this analogy is ill chosen, for it is misleading, since even though quantity be related to quality within value, it is not at the level proper of value, but at the much lower level of combined theoretical and empirical activity that quantity is being submitted to mensurations, for instance in music where rules of a proposed counterpoint and laws of an assumed harmony do combine, within commensurability, parameters like the notes of scales, the shifts produced by sharps and flats, *forte*'s and *piano*'s, *crescendi*, *ralentendi* and so on ... respecting of course the main difference that whereas the promotion of truth resides in the establishment of universals as ideas thought to be universally valid among particulars collected in classes, the promotion of beauty rests upon the creation of particulars as individuals thought to be faithful illustrations of ideas (themselves held for universal). And then the universals (not the particulars, even though they may have been used in 'verification') are in the former case the bearers of Truth, whereas the particulars are in the latter case the bearers of Beauty.

And finally, as far as goodness is concerned, neither of the two preceding reasonings apply, since the promotion of the good does neither consist in establishing universals from available (classes of) particulars, nor result from the creation of particulars out of the awareness of ideas. Much rather, it consists in the 'realization', in an act (often called 'actualization'), of that which is common to both a universal and a particular and makes, that the universal and the particular coincide, a coincidence which cannot become a fact without there being two 'factors': one giving and the other receiving the 'benefit' (from Latin: *bene-facere*) of that coincidence.

And this is why the distinct promotions of the three cardinal values necessarily rest on three respective, distinct, attitudes within the modality of judgment or judicative modality: as opinion, estimate, critical faculty, discernment, good sense and all such promoting faculties of thought. They are: the objective, the subjective, and the conjective modes or 'attitudes' (derived however from *aptitudinem* and not from a non-existing Latin word *'actitudinem'*), in the very etymological meaning of these three adjectives, objectivity having the searched, *true* idea be found to assume the common lead of the known particulars, - subjectivity having the produced, will-be *beautiful* particulars be found to realize the idea as individual things, - conjectivity having the community of aims be the very nature of *good* deeds.[1]

[1] The circumstance that evil, uggliness, untruth are encountered in World, poses a specific question within the problematics of God, World and Time, because God is held to be Perfect, whereas these three antivalues are held to embody the opposite

Above statements might support a traditional opinion among Greek philosophers tending to assert the supremacy of a highest good, τὸ πάνυ ἀγαθόν and even the Aristotelean idea of a happiness (εὐδαιμονία) at possessing it (cf. the *Nicomachian Ethics,* e.g. Book I, chap.VII, no. 9), and generally speaking the linguistic use of the adverb εὖ (= without exception) as a prefix in various words designating high(est) values (also τὸ εὖ: the good, right, justice), letting however truth (ἡ ἀλήθεια: also good faith, - notice the adjective good in that translation), - and even beauty (τὸ κάλλος) hold a second, even though most important, rank among the highest values.

Whether Christianity has changed such hierarchic views must probably be negated, not because of a different philosophical approach or reasoning however, since Christianity was the foundation of a religion, not of a philosophic system. Yet something happened there in the attachment of such 'value' as the supreme good to Divine origin. Of course ARISTOTLE had already suggested it (ibid. chap. VI, 3ss. where the good is said to possess as many categories as the To-be and must as a substance be called God and Wisdom). But ARISTOTLE's work is philosophy, not Holy Scripture. Christianity seems, to begin with at least, to have ignored beauty and truth (in the sense envisaged here). And this is in agreement with the fact that the Jews among whom Christianity was borne (and born) have always shown interest for ethics, for sure, but neither for mathematics nor

of perfection. This is the subject-matter of the paper by A. MERCIER entitled *The Problem of the Imperfection of a World, itself created by a Perfect God*, publ. in Foundations of Physics, Vol. 22, 1992, pp. 205-219, and dedicated to KARL POPPER on his 90th birthday. The answer given there is that, since God cannot fetch the material of which World is made 'somewhere else' (i.e. outside World itself), for then World would already exist 'there', He must tear it along Time from Himself, thereby violating His own Perfection (or 'Symmetry'), necessarily injecting imperfection into the To-come, since He cannot as the (only) Perfect Being part with this Perfection which is 'absolute and infinite', whence the finiteness and imperfection of World in Time.

for æsthetics as the leading activities towards the promotion of Value. And we should not forget, that within Western culture as it has been moulded not only by Greek, but also by a Christian tradition itself leaning on Jewish tradition, the notion of value is much stronger attached to the pursuit of the good than to the promotion of either beauty or truth.

This was later corrected - within religious thought - first within Islam which declared God Lover of all three cardinal values, and in the later course of the centuries of course, when philosophy and especially logics gained increasing importance in scholastics, so that even a certain shift may have happened: Truth became more and more respected, be it a Divine truth, or one of philosophical nature; and in our days, the (not only atheistic) tendency is widely spread to give 'truth' (whatever is meant as such) the lead, yet not (or no more) a Divine truth, but a very secular truth, concerned with World as such, however not a truth 'of' World, but truth 'within' World 'through the means of discourse'. Whence the flourishing of sciences taking over even the consideration of the good: the 'social sciences', which easily forget, that not only Man is a sociable being (cf. again ARISTOTLE, ibid. I, VII, 6.), but most, if not all living beings, e.g. trees, are sociable, else we could never eat an apricot if not two apricot trees would grow sufficiently near each other for bees to fructify the blossoms from one to the other tree. However, current ecology is not of much help in clearing up the relationship between God, World and Time with special reference to the axiological problematics of Truth and the Good, though precisely this should be its purpose, - to say nothing of beauty, unless it is made a technical discipline[2] at the encounter of science and law i.e., concerned with the search of both truth and the good at a time. Such example

[2] **Technique is *always* at the encounter of two disciplines pursuing each a different (cardinal) value (as has been explained e.g. at the Meeting 1991 in Prague of the International Institute of Philosophy, this being the reason why the neologism 'technology' used in our days is falsely constructed).**

as well as others suggest very strongly, that the supreme value does not simply coincide with one single cardinal value like the good alone, but with the promotion of all three in One loftiness that reaches the highest possible within World: the sublime i.e., reaching its lintel which parts it from divine heavens, hence at the very source of the To-come.

In Greek, λογικὸς is not only 'relative to discourse', but also 'endowed with reason'. If it be put in the plural, *logics* will designate the forms of discourse and their entanglement that are precisely endowed with reason in a way that makes discourse persuasive. And that is the way to be followed and the condition to be fulfilled.

Then ποιητικός means capable of doing, of creating, of producing. (For ARISTOTLE, τά αἴτία ποιητικά were the efficient causes; for PLATO, ἡ ποιητική was the art of the poet.) Put in the plural, poietics will designate the forms which in art are endowed with the power to achieve. Hence *poietics* is the way to be followed in creation.

Finally, τό δέον is what must be done i.e., duty (from δεῖ: must). Hence, by forging the word *deontics*, it will designate the forms or rules of precisely the obligation implied by the verb must. And that is the way to be followed in (moral) inter-course (not just dis-course).

Logics is therefore not itself the very discourse which discloses truth, nor is poietics the artistic creation as such, nor is finally deontics the mere good action performed in morals. Each of them exercises the control for these three activities not to deviate from their proper aim and design. Of course, to teach them can also be a matter of discourse, yet it is then, in analogy with the French designation for philosopohy as *une réflexion seconde,* say *un discours second,* which is however best learnt by its practise in each of these cases.

So far, we have referred these considerations to the domain of commensurability. If we want to transpose them to the incommensurable, where metaphysics and the corresponding stages of thought are necessary, we put at the level of pure form other names to designate these exercises of control, namely: dialectics along the search of truth as the highest form of discussion as if that were 'the way the gods talk with men' (PLATO), then theopoiïa (the form to be taken by the creation of an adequate image of God) along the promotion of beauty, and εὐχή (from εὖ as already mentioned) which implies humility as of a prayer addressed to the divinity.

Underneath the level of pure form stands the level of pure power, of pure will, and of pure duty.

Within commensurability, and as far as truth is concerned, it implies mathematics; within incommensurability, apodictics. This rests on the fact that, while logics is, popularly said, the theory of proof, mathematics is the execution of proofs; and, within incommensurability, while dialectics is the theory of metaphysical discourse, apodictics is its metaphysical clear demonstration proper. And whereas the mathematician (in Greece the 'geometer') could be said to be the contemplator of truth i.e., he who dwells within the mathematical domain (ARISTOTLE, op.cit.), the æsthetician is the contemplator of beauty and the ethician (this word is not used as such in Greek) is the contemplator of the good, admitting successively of a metaphysical manifestation (ἀπόδειξις) of the divine, an intuition or vision (ὄψις) of God: θεοπτεία, and of a metaphysical ordination of divine nature (εὐταξία).

It should by now be quite clear to the Reader that mathematics are not to be classified among the sciences proper: Mathematics do not constitute a science, because they do not actualize knowledge (Latin: *scientia*); they constitute a power (*potentia*: also faculty, ability, efficiency, beside other senses as germane to political power, which then do not apply to mathematics). Political power is again and again usurped, but mathematics is never an usurped power, the reason why it is pure power. The corresponding verb is *can,* for there is not in English a verb as derived from the Latin like the French *pouvoir* having also yielded the substantive noun identified with it; German has *können*, whereas it says *wissen* for to know; hence, beside *Wissenschaft* for science, one could in German forge *Könnenschaft* for the power as embodied by mathematics.

A parallel commentary holds for the pursuits of both beauty and the good. That is why æsthetics and ethics stand at the same level as mathematics, yet not as pure power, but as pure will (*volentia* from *volere*) and of pure duty (unfortunately, there is no corresponding Latin word as derived from *debere*).

Together with to be and to have, and a few other perhaps, the verbs: *can*, *will* and a third one that is to be borrowed from another etymological root, viz. *must* (German: *müssen*, for there is not a verb to 'due', and to 'owe' has another meaning), belong to the most important ones in language and are at the same time found to figure among the most irregular verbs like in most indo-european languages. It is therefore not astonishing, that they are at the outcome of the three cardinal values, the 'to-be' and the 'to-have' not being regular 'values' but embodying thoughts of an altogether comparable weight. These remarks are applicable not only in the domain of mere commensurability, they hold as well for metaphysics, for poetry, and for personal morals. Only at that level does one encounter the necessary combination of the theoretical with the empirical towards the ascertainment and confirmation of truth, of the compositional with the executional towards the yield and illumination of beauty, of the critical with the performative towards the release and actualization of the good. This is also valid within the incommensurable, where the adequation can be specified to obtain as a 'meditation' or exercise of the mind involving the acknowledgment of death (the surest certitude involving truth) as founding freedom (the most conspicuous token of life involving beauty) within the reach of love (the closest intimacy involving the - supreme - good):

μελέτη περί θανάτου καί ἐλευθερίας καὶ (θεο-)φιλίας.

Chapter X. Theodicy, or the Metaphysical Vindication of God i.e., the Knowledge by Reason and Experience of the Divine Person

1. The Reasons Why a Vindication of God is Requested

It is possible that such as call themselves atheists, after reading the former chapters, will exclaim: This can only be nonsense, since it involves the idea of God, while there is no god, for while one can speak or write about e.g. World and Time, one cannot do so about God, World and Time.

The clearest retort that can be given is that there have never been attempts made at *proving* that there is no god, for the simple reason that the proof of the negation of being is not feasable since there can be no corresponding experience, and therefore the assertion is at most either a hypothesis, or a belief, but a gratuitous hypothesis like that does admit of no verification, and a belief does as such not enter philosophic reasoning, - whereas there have been attempts made at showing by reason that there is God as confirmed by very general human experience.

We have already excluded as non-sensical the eventuality to talk about a would-be 'existence' of God, since existence is a property which cannot be attributed to Him owing to the circumstance that He is anterior to existence, so that in classical texts which do use it, the phrase 'the existence of' in front of the word God must simply be dropped or at most be replaced by 'the being of'. The most correct procedure would be, even the only reasonable is, to talk and write about 'proofs of God', but never about proofs of His existence.

In that sense and consequently, arguments in favour of the assertion that 'there is' God may be looked for, and then put forward, eventually to be submitted to a critique under the condition that such critique is up to its object, never using 'belief' as an argument.

Also to say for example that no case has ever been known that He could be perceived by one of our senses, e.g. by sight, is inacceptable, since the senses do not communicate but the 'existence' of *finite* 'things' which must be commensurable to Man, yet things are parts of World and change in Time.

Therefore a 'theory' about God cannot be put forward to be 'verified' by experimental means in the (same) manner scientific theories (of gravitation, of electromagnetism, of evolution etc.) are being put forward about Nature and tested towards veri- and/or falsification. And even though the three-fold junction of God, World and Time reminds of the form as postulated in our theory of knowledge and thought to make it at all possible to understand each term through its junction with the two others does not suffice to make it possible to verify any-*thing* about Him; especially, - but not only, - we can never dispose of the totality of World or of Time in any of our experiments or observations, or in our experience generally speaking. There is indeed but one single, very exceptional case of our experiencing God, namely the *authentic* mystical experience, which however cannot be prepared for the sake of verification (or falsification), but is found to be *granted* to subjects without their asking for it (even among yogis or dervishes), so that no any human being can 'claim' to intentionally take a share in it and be therefore in a position to eventually feel justified in 'concluding' that 'there is no god', - while objectors practically always express their doubt by saying that "they simply do not *believe* in Him," which of course is an inacceptable conclusion, since belief does not belong to any *reasonable* activity of the mind.

Therefore, since there are individual cases of mystic experience of such exceptionality which cannot honestly be refuted, - and indeed exceptional not in number, but in obviousness, - to deny it e.g. on would-be medical grounds and eventually destroy it by so-called psychiatric methods and artificial means of chemical or other remedial therapeutical intervention involves the most serious possible reponsibility, - a reponsibility which can indeed be assumed by ***nobody at all,*** since it amounts to what might and eventually must be described as crime which does not merely concern the physical existence, but its metaphysical surpassing beyond the bodily nature of Man.

Yet this also leads beyond reason, since it concerns mystics and does not allow to consider mystical experience as a proof of God 'by reason' i.e., within the field of mere philosophy including metaphysics proper, nor within the field of religion proper as complementary to philosophy, since the contemplative modality is neither part of philosophical, nor part of religious thought, but the *link* between the two, as has been explained further above. And the Mystic experience, if involved as a means of verification or falsification, must be accepted as an experience which overhangs the objective, subjective, conjective and even religious attitudes. But if ever there should be no god, even the metaphysical dimension itself which has been found, for Man, to correspond to the incommensurable, collapses, so much so that there would be no situation of incommensurability, and that atheists could not even claim to enter the domain of metaphysics: To them, there can be no such domain.[1]

[1]Nor could there be, to them, any hope to actually enter the domains of either poetry, or morals of the self!

Conversely, if we claim that there is a field of thought to be dedicated to metaphysics (as well as the corresponding activities along the promotion of beauty in the subjective, and of the good in the conjective, attitudes), we should undertake the task to justify the claim, that there is indeed the Incommensurable: both quantity and quality of a Value which is not attached to commensurable *things of World in Time,* but to the Being Who is at the source of both the To-come and the Creation of World. Within the Christian tradition, this very need has been strongly felt as soon as thinkers have started to reflect philosophically besides religiously i.e., to devote themselves to that aspect of thought which is complementary to pure religion, and a tradition has thus been established to devote oneself to the problem of the possibility and performance of proofs of God (whereas the problem of the possibility and performance of proofs of World - that is through the sciences - did not retain their attention).

When the Instigators of the Third World Congress of Christian Philosophy prepared the same to be held in 1990 in Quito (Ecuador) on the general theme of 'Present-day Atheism and Divine Transcendence', the Author of the present Essay was requested to deliver a Report concerned with the problematics of the Proof of God Today, within a Section devoted to the alternative of 'Experience of God or Proof of God?' This report has been presented in Spanish and is reproduced in the Proceedings of the Congress (publ. by Ed. de la Pontifica Univ. Cat. del Ecuador, cf. Vol. 2: *El Tema de Dios, un Tema Antropológico,* pp. 213-230). Since it contains as a matter of fact the very argumentation just requested further above, it seems appropriate to reproduce it in a slightly revised English version adapted to the purpose of the present enterprise.

2. The Problem of the Proof of God, Today

(1) It is usual to classify (former) proofs of God according to various kinds as follows (we summarize each kind of argument).

(i) Ontological proof: "God is the Perfect Being. Existence (as meant in those proofs) is a perfection. Consequently God exists."

Comment: This rests on the idea of existence as a perfection, which could be contested, for how do we know that existence is a perfection? Moreover, according to what has been said further above, we should avoid the attribution of existence to God. Therefore, such an argument shall not retain our attention. Furthermore, ontological is perhaps not the right adjective to be used here, for one is either referring to that which possesses existence: τὸ ὄν, which does not concern God, or to the very To-Be: God, ὁ Ὢν Himself (the genitive of which is , 'tis true, Ὄντος), but then one is concerned with theology properly speaking.

(ii) Cosmological proof: "If there exists anything at all, there must exist a necessary being. Since I exist, that necessary being: God, exists."

Comment: Again the use of 'to exist'. and the problem arises to modify this formulatioin in order to avoid the attribution of existence to God.

(iii) Teleological proof: "World is manifestly filled with designs which assume the rôle of final causes, the disposition and nobleness of which reveal an infinite wisdom which the worldly things cannot embody otherwise than in a contingent way. Consequently, there is a prime cause altogether almighty, infinitely intelligent and perfectly free: This is God."

Comment: This avoids the reference to existence and renders that type of proof much more satisfactory. Especially the almightiness of God makes out of Him the (only) possible Creator of World (and Time). At the same time, the teleological nature of the proof is secondary, since it makes God appear as the singular primary Cause of all there is (and exists).

(iv) In the 20th Century which has seen the appearance of formal logics, a kind of proof has also been looked for which could be called a *logical proof.*

Comment: Of course, any proof must be logically conceived. What is new in this case is the idea that a proof of God could be - and, in the mind of contemporary logicians like e.g. BOCHENSKI, should be - exclusively based on formal logic (cp. the Chart on p. 69) as performed at the level of *pure form* not only above the levels of theoretical and empirical research, but above the level of pure power (i.e. of mathematics), in other words, since it is a question relevant to the incommensurable, within the application of a formal logic as adapted to Dialectics. This rests on the idea shared by formal logicians of our time that logic is not a matter of convention, but that it embodies truth itself and shall therefore be used to prove that there is a theory and only one theory necessarily true, which explains the class of all facts and will never be refuted. So much so that this very theory coincides with the major of all possible facts and consequently coincides with God as endowed with immortality.

Such types of proofs, above all the ancient ones, to which are particularly attached the names of ANSELM, THOMAS, DESCARTES, MALEBRANCHE, ... have been formulated by

Christian thinkers who seem to have felt the intellectual need to propose such proofs. Yet the very thought that it makes sense to speak about God and to develop about Him a valid *argumentation* goes back to ARISTOTLE, to PLATO, to PYTHAGORAS, ... to the *Vedas*.

KANT, who spent his life criticizing everything, criticized the proofs known to him, and proposed his own. That is a procedure dear to certain prolific authors who begin by demolishing the doctrins of their predecessors in order to propose their own in their turn. Perhaps this remark could be made around our own comments. Yet actually, what we do, is generally speaking to claim that a proof has to be a proof of God and not a proof of His falsely assumed 'existence'.

(2) In our days, only mathematical proofs are considered rigorous. But it was found more difficult than ever to strictly distinguish *logics* as converted into *formal* logics, from *mathematics*. We have chosen to designate as logics the 'doctrine of proof', while mathematics are constituted by the evergrowing whole of the (strictly formal) proofs which accumulate in the course of history. Yet under such a point of view, only the *formal*, but not the *substantial* proofs are involved, which means that they do not discover truth as a value, but only the perfect forms of a possible discourse of which it is hoped that it will serve for the establishment of truth. This, by the way, is the only sign of the actualization of perfection within World, and it has its parallel along the promotion of the two other cardinal values, beauty and the good, yet in a concealed manner which is very difficult to make perceptible. Moreover, the proofs in question here aren't but the proofs about sets of elements the determinations of which are finiteness and multiplicity. Especially, that finiteness and that multiplicity seem to be inherent to

World as constituted by things, in Latin: *res*, which define that which is commonly called Reality.

The elements as spoken of by mathematics are never, themselves, such things and hence do not constitute reality. Opposed to the latter which are beings of fact, these elements are called beings of reason or fictitious beings: a mathematical point, for instance, is never a grain of sand nor is it a flea or a star; a geometrical angle is never the corner of a wall nor is it the extremity of a knife; a group of symmetries is not an optical reflection in a mirror nor the blade of a tree with its nervures. The question to know whether God is a being of fact, or a being of reason, will not admit of an answer unless as established by the specific contents of the proof which is proposed of Him. Yet we can hardly expect to get either of the two possible answers, for the plain reason that He is anterior to both reason and fact!

The theorems of mathematics are rigorously established, which means: logically demonstrated from systems of axioms considered non-contradictory.

Yet it is of clear understanding today that axioms are not 'true' in the sense where to be true requires the coincidence of form and fact, and consequently mathematics are neither true, nor false of course; they are never 'verified,' - i.e. made true - by any fact; hence they do not possess the status of a science - in Latin: *scientia* -; what they possess is the status of a (pure) power - in Latin: *potentia* -, and what in the philosophy of mathematics is sometimes called 'platonism', consisting in the assertion of the coincidence of World as real with mathematics in their totality, is a proposition devoid of sense and significance not only because it is impossible to confirm or contradict it, but above all because it amounts to an identification of fact with fiction.

(3) The circumstance that mathematics are found to be the most acute and efficient formal instrument in the development of all science concerned with reality appears as an 'epistemological mystery'. It is a mystery in particular for the reason that the mathematical elements as used in scientific explanation never do identically coincide with the elements of Reality: Celestial bodies are assimilated to material points, what they manifestly are not; small quantities of matter are identified with infinitesimals of mathematical analysis, and one proceeds to still more serious identifications without being in a position to justify them by anything but by the success which results *a posteriori*. Physics in its entirety is the most conspicuous illustration of this.

DESCARTES inaugurated it as a postulate of a well made science, and NEWTON introduced it as a final procedure; physics has never given the lie to it since. Moreover, this Cartesian postulate is not meant to be applicable to anything else than World itself, however big it may be, and to its parts of course, yet not to that *which transcends the world.*

Only in the second half of the 20th Century was the idea of a formal logical proof extended to the Capital Being, as far as is known to the Author of this Essay, by the former logician of Yale University FREDERIC B. FITCH[2]. Yet, when this is looked at closely, one notices that the Capital Being thus implied is not identical with a God who transcends World and Time but rather with a World of Logic in its entirety in the sense that that entirety would be 'more' than the reunion of its parts. Then we can ask whether this is the God who would have created World and Time.

[2]See e.g. *Combinatory Logic and the Theory of Ingression*, in *Philosophes critiques d'eux-mêmes - Philosophers on Their Own Work - Philosophische Selbstbetrachtungen* (Peter Lang ed., Bern-Frankfurt/M.-New York, Vol. 13, 1986, pp.67-89).

Now, in its factual reality, World is found endowed with both finitude (even though it may eventually be necessary to attribute to it huge dimensions e.g. in a Riemannian geometrical representation), and multiplicity. And what is more, there are not World *and* Time as separate. For the prime evidence about World is that it changes, and the prime possible assertion about Time is that Time cannot be numbered without the help of World and its parts. In particular, reality, in its innate finitude and multiplicity, manifests two phenomena which appear punctual within their eminently temporal nature: *birth*, and *death* or, if we wish to necesarily avoid evocating life in its biological sense, the apparition and disappearance of entities, e.g. the sequence of particles or other microphysical entities which within cosmology might have entered World according to the big bang hypothesis, or which may disappear as a consequence of mutual collisions, or may come out of radioactive disintegrations. So much so that it looks as if World were the 'subject-object' of appearance and disappearance, the former of which has suggested to all peoples of the Earth the myths of creation of World, for which for instance the Greek made the *demiurge* responsible, or as if, in the Biblical tradition, yet going beyond it, "God had separated World from Himself", which assumes a Being as *capital, all-mighty and free,* Author of the said creation, whereas none of the myths does assert the disappearance of World as such, unless we make a reference to the idea contained in eschatological doctrines of Christianity.

One can criticize as anthropomorphic the argument based on a demiurge. What is certain is, that it is impossible to yield an explanation for the *existence* of World without recourse to an argument which would evoke a genesis. Contrarily to the simple-minded opinion of many rather candid scientists, the recent theory of the big bang does not render easier that explanation, since it must postulate the previous presence of that which LEMAÎTRE had called a 'primitive atom' of which the big bang would be nothing but the radioactive explosion, and about which it must be assumed, that it was already there, without knowing what the adverb 'there' means, while the adverb 'already' specifies that it was there within time, hence that it 'existed' before the big bang, i.e. before the existence of both World and Time; and this is self-contradictory.

One must, indeed, be conscious of the fact that under such conditions the very word 'existence' does apply to nothing but to the things of World which are in Time as well as to World in its entirety which is accompanied by Time since the primitive atom was 'there', if World did indeed proceed from it. In other words: *Existence is the to-be - or being - in time.* Consequently, existence cannot be an attribute of the demiurge (still less of God), unless that demiurge had been himself created - as is for instance suggested in the teaching of Hinduism, where the creating god, BRAHMA, also preserver and destructor of World in the respective persons of VISHNU and SHIVA with whom he constitutes the trinity called Trimurti, is himself the emanation of the ***brahman*** i.e., of the principle of all things. (And that sounds pretty much like God detaching World and Time from himself.)

Then, if a philosopher grapples with the problem of knowing whether God (as being at least the demiurge) can be proved, the said philosopher must abstain from attributing existence to the god he proves, since God is not within Time, within a time which He could not have detached from Himself together with World which He should have fetched there where it already was present in a primitive form. Hence it is absurd to speak about 'proofs of God's existence'; one can only and plainly speak about *proofs of God ...* unless one intends to say that God *incarnates Himself;* but that is quite a different problematics (which is dealt with in Hinduism in the idea of the avatars of Vishnu, and in Christianity in the idea of Christ).

(4) The Platonic adage Ἀεὶ ὁ θεὸς γεωμετρεῖ itself, according to which the demiurge always acts as a geometer i.e., as a mathematician, has apparently - but only apparently - an anthropomorphic touch. The same is still truer of assertions which make out of him a great watch-maker. For that reason, such assertions cannot easily be held as arguments or proofs of God, since in addition to their would-be anthropomorphism which might make them somehow blasphemous, they rather concern World if World is His creation, and not God Himself if He does not seem to create World.[3]

---

[3]Hence, one should be very careful about the use of an 'anthropomorphic argument' uttered against any attempt at talking reasonably about God: The idea behind such descriptions as attempting, like the above quoted Platonic adage, to give a picture of (the) God, is not to make Him resemblant to Man; quite the contrary: it aims at calling the attention upon the fact that Man appears to the philosopher as 'endowed' with capacities which can very well be described as reminding of much superior capacities owned by the divinity but which cannot be adequately described by human language, since what is merely human will never suffice to serve as a means identical with the means at the disposal of God. Hence, strictly speaking, any such anthropomorphic argument as used by atheists in order to discard any discourse as used in the attempt to argue about - and often even in favour of - the working of God, is a *fallacy*, even though historically speaking it may well be, that such and such attribution of characters to divinities within polytheisms of various kinds may have arisen as anthropomorphic designs. For indeed, a saying like above Platonic adage, which asserts that God *always* acts as a geometer (read: as a mathematician, whose 'mastery' over mathematics is total, universal, inexhaustible, insuperable ...), does in no way ascribe to God a human quality; much rather, it asserts that any human mathematician will 'never' (in opposition to the "ἀεὶ": 'always') be capable of doing all what he does with mathematical perfection as God does. And this implies that Man has only a resemblance to God, yet a resemblance which makes him an imperfect creature as compared with God's perfection.

The same thought, yet told differently, is implied in the passage of *Genesis* (cf. *Old Testament*) which says that Man was made as an image (evidently less perfect) of God. This could at most be used as a *theomorphic* argument within an anthropology of non purely scientific nature, but in any case *not* as an anthropomorphic argument within a theology of any kind. Will-be atheists who assert the latter simply fall into a gross fallacy.

Moreover, however impressive be the geometrical aspect of World, its temporal aspect, - suggested by the image of the watch-maker-god, - is perhaps still more important. For one should not forget that every well elaborated and well verified scientific theory amounts exactly to a *conceptualization* of Time. For instance, Newtonian mechanics defines a Newtonian time, Einsteinian mechanics defines an Einsteinian time, thermodynamics defines a thermodynamical time ... and the Darwinian theory defines a time of the evolution of species, and so on. The whole of the sciences is therefore found to constitute a progressive way of a common conception of Time and World and puts for us the question, to know whether there is a valid science which will allow to *make abstraction of both World and Time* in order to validly extend it to *that, which transcends World and Time.* If such a science is possible, *then it is metaphysics;* therefore it will not be the first, but rather be the last of the sciences, as situated at the exit of the undertakings of objective knowledge, - a *science sui generis* however (but not to be confused with mathematics!), since instead of yielding on its part one more, or a more elaborate and more general conceptualization of Time, its aim is actually to do the opposite, namely to establish the conceptualization of *No-Time* (and of No-World, though the latter should in no way be confused with nothingness).

For all that, one shall clearly understand that if a proof of God can succeed, it must happen also with radical transcendance beyond the finitude and multiplicity of World. And since mathematics (to which formal logic is to be added), inspite of the enormous success of the theory of sets, have never been able to get beyond the first conceptual grade of infinity called the transfinite, which is and remains after all a kind of finiteness, since it always is feasable to surpass it in thought, - transfiniteness which, consequently, mathematics are unable to transcend towards a radical infiniteness, - it becomes necessary for the thinker to have recourse to a discursive method of an order superior to that of mathematics (and of formal logics) if he is to hope to succeed in elaborating proofs as adequate to the 'object' in view i.e., God.

And yet we should rather not maintain that, in analogy with metaphysics, this transmathematical method ought to be designated as meta-mathematics, for the reason that it as such would rest on a delusive analogy. ARISTOTLE did not call metaphysics his tentative to speak about God; he called it *theology.* And that is what is precisely concerned here, under the condition that the word be not confused with what would later be considered theology: as Christian, Moslem, or other, thought of religious nature as hermeneutics i.e., interpretation (supposedly trustworthy) of the holy scriptures. For we are not here involved in religion. Religion inspires confidence - in Greek πίστις - which in its own way is a testimony susceptible to be looked upon as a 'proof', yet then as a proof of an other nature, but not as the kind implied in the quest of those as called proofs of God which, in their turn, are of philosophical nature. But since the name of theology has been derived from religious (and not from philosophical) considerations, it is difficult to restitute it in its original sense which would have been the only valid one for metaphysics: as science *sui generis* of the proofs of God.

In parallel with this, one must invent another name for the discursive method capable of attacking the questions whose object transcends World and Time. Our opinion is that this is *dialectics,* the distinctive character of which, in contrast with mathematics which rest on deduction, is to allow a mastership over that which intelligence has in each step of its course been able to assert.

However, including the possession of that dialectics, this

would not suffice if we did not also possess a means to establish the proofs looked for not only within the total abstraction of dialectical discourse, but also in its reference to an *experience* of that which transcends the reality of both World and Time. And that is perhaps the matter in regard of which opinions diverge most.

(5) One could say, - as JOANNIS THEODORACOPOULOS used to - that PLATO has established the dialectics of the To-be and especially that of the Major Being even though he had precursors among Vedic thought from which, most probably, PYTHAGORAS appears as on an intermediary stage. Yet whereas the *Vedas* made in those old days constant reference to a transreality which they personified in such and such divinity, and whereas PYTHAGORAS caught within 'theory' a glimpse of experiential as well as of speculative procedure towards Divine Being, the idea that it suffices to argue within abstraction about God in order to prove Him has by and by established itself in the course of times which saw the coming of successors, especially ARISTOTLE. In this way, a 'speculative philosophy' came forth as a kind of metaphysics devoid of the experiential support required by every well-established science. For that reason, it does not fulfil its proper task. And yet, such a blame does apply neither to PLOTINUS nor to S. AUGUSTINE. For it seems, to begin with, correct to say that PLOTINUS appealed to *mystics* as the experience assumed capable to serve for the verification of the dialectical apprehension of the To-Be and that in conformity with THEODORACOPOULOS' assertion we owe to PLOTINUS (but not to ARISTOTLE) the metaphysics proper of the To-Be; and furthermore that S. AUGUSTINE - in contradistinction to a platonism as is usually attributed to him - is to be declared the founder of *existential philosophy:* however not as atheistic (like in our days), but as authentically based, in the last instance, on the transcendence of the experience of that which is beyond or outside both Time and World,

the two later great adepts of which will be: PASCAL in modern times, and BERDIAEFF within contemporary philosophy (not mentioning among them GABRIEL MARCEL whom we shall not refer to for the reason that his problem was the problem of faith, not of proof, and that he always refused to be designated as an existentialist). For, after all, that which the proof of anything is expected to be established, - if not within mere mathematics with their theorems, - is *truth.* In particular, a proof of God is only conceivable as an authentic Truth i.e., some kind of view upon God which could have been falsified or invalidated by the experience we make of Him, yet which has indeed been confirmed: verified. And saying so is compromising oneself quite seriously, since it amounts to applying to the disclosure of truth about God POPPER's criterium about falsifiability which allows to qualify as true a scientific theory, whether physical or (beyond POPPER's own intention) metaphysical, as long as it has not been falsified but found adequate to its object.

The burning question to be put here would be to know which is the nature *sui generis* of the said *metaphysical experience.*

Now, on the side of World, the proofs of which physics and the other sciences (said to be positive) give of it bear the double aspect which has been found necessary for truth in order to perform the actual proof. It is subsumed in S. THOMAS' formula of *adæquatio intellectus et rei:* for a conclusion to be true, the theory put forward about reality must be corroborated by the facts with sufficient approximation. Mathematics on their own side are never corroborated by facts; they are never 'verified' - nor are they falsified either, of course. And if we pretended to be satisfied by some transmathematics, they would not either suffice for a proof of God.

ARISTOTLE's physics has been found wrong since it is falsified by observational facts. DESCARTES' physics had the same fate, as far as his theory of vortices is concerned. Have the respective metaphysics of these two authors been found true, or perhaps wrong? This question, as far as is known to us, was never put in that particular form. Anyhow, one cannot forbear to guess in the later development of philosophy a progressive desinterestedness towards metaphysics, whatever the latter be, which even led to a disapproval or a condemnation, depending on the case at hand. At the same time, that which could have remained in relationship with metaphysics had decayed, so much so that, at the opening of the 20th Century, there was, properly speaking, not left much of metaphysics if, in any case - and *toutes choses égales d'ailleurs* -, the correct analogy resides in that metaphysics be theology in precisely the sense in which, within another order of thought, biology is the science of life, or physics the science of nature (which 'could' if not 'should' be called physio-logy, from *phusis*).

Later, especially in our days, - say after the decay of positivism, - a renewal of metaphysics became noticeable. We should like to assert that if that renewal is to be authentic, metaphysics must succeed in arising as theology i.e.: in having as central task to 'prove God' just as physics, like the other positive sciences, has the task to 'prove World', that is: make the truths about World explicit, scientific theories being verified, not falsified. Yet if, a few lines further above, we took the care to add *"toutes choses égales d'ailleurs"*, the reason for it was that there is between World and God an *incommensurability* which stems from the fact that, whereas all the human measures are quite adapted to measure World up to its cosmic quasi-immensity, it is not immediately certain that Man has at his disposal the experiential means to 'measure' God: to perform the experience of God.

Now, only a metaphysical theory as corroborated by an experience of God - a metaphysical experience, as well - can constitute a respectable science.

Under such conditions, the question arises to find out whether one can, and even whether one ought to, approach God - both theoretically and by experience - in an *objective* manner. If the contention is raised that God Himself cannot be made 'the object' of such an approach, such objection can be met by requiring that this be proved to begin with.

Within Plotinian thought, the approach cannot be said to be objective, since in it mystics assume the rôle of experience. If one requires that the corroboration of metaphysics as a science be objective, the Plotinian procedure must be rejected. As far as PASCAL is concerned, it is less easy to be specific, since to him the situation was for one part that of a fear which he felt towards *"le silence éternel de ces espaces infinis"*[4], and for another part the mediation experienced in JESUS-CHRIST[5]. However, one feels there something of a physico-cosmological, hence objective, nature of the formidable experience of the infinite spaces, as well as something deserving the description as objectively mundified in the experience of the presence of CHRIST within messianic time. As such, the Pascalian step is that of an objective mode of thought. Let us not forget, that PASCAL has been a great mathematician and a shrewd physicist. In BERDIAEFF, we have to reckon with the experience of the 'reign of the Spirit'. We shall come back to it further down.

(6) The objective attitude is that of a subject who intends to project the object of his consideration onto a plane where he can make of it both a discursive and a substantial analysis, his aim being to insure the biunivocal adæquation of

---

[4] *Pensée* no 184, cf. Edition TOURNEUR.
[5] Ibid. no 176.

the two analyses. (We must insist here upon the circumstance that to explain objectivity as the consensus among all is undefendable, for the simple reason that a consensus cannot obtain if not among experts as well as because it is applicable to non-objective enterprises as well.) Thus, we have here to deal with a sort of chemistry, and we remember that the word 'chemistry' is derived from the old name given to the black earth of Egypt where the antique precursors of alchemy resided. Well: does God dwell within the black earth of a country which is not in this World, and which then ought to be localized? Evidently no. It is clear that, *sensu stricto,* God is not to be encountered in the heavens of the astronomers, behind the clouds or the galaxies, since he is outside both World and Time which He, as we have learnt, has detached from himself: for Him to be God and remain God, He must *abscond* and keep absconded. But since He is not hidden at some place of World, for we'd have found Him there long ago, He cannot but abscond *behind Time,* yet not in the past of Time, since He would be 'no more', much rather and necessarily *in the future* i.e., in the To-come. And only thus, by the way, can we experience Him, an experience which is precisely mediated by Time itself, a time which comes and persists in coming to meet us, loaded as it is of all what we are able to experience as possessing the value of the sacred. According to PASCAL, precisely, the most manifest mediation, the authentic one even, is that of the messianic time as embodied in JESUS-CHRIST. In that respect, and understood as has just been explained, that experience is objective: *transobjective* if you insist upon making the difference implied in the phrase *"toutes choses égales d'ailleurs".*

(7) The more we look for entries that open a way to introduce ourselves into metaphysics by however respecting both the theoretical and the experiential exigencies, the closest we get to the conclusion that such introduction requires a dis-

cursive instrument superior to that of the ordinary sciences as well as an experience of temporality coming from beyond Time itself, for the very simple but deep reason that between World and God, there is an incommensurability which the usual procedures of natural science do not allow to master.

In the ordinary sciences, the introduction of concepts which it will be the task to submit to experience of facts happens with the help of what are called *implicit* definitions. One knows that the notion of definition has given rise to controversies, the history of which goes back to Port-Royal, if not to ARISTOTLE. It seems that such controversy originates from the fact that it is doubtful whether a universal and rigorous definition can be given of definition itself, since it would amount to enunciate a proposition speaking about itself, and it is well known that to do so implies difficulties which RUDOLF CARNAP has treated in his *Logical Syntax of Language*. The mathematician HILBERT, himself a very good philosopher, has helped epistemology along an important step when he explained that beside explicit definitions, there are implicit ones richer in consequences because they imply several notions at a time, the distinct senses of which become clearer and clearer in the prolonged praxis of a system of talk as applied to experience, especially in physics and in science in general.

Yet after all, the idea of implicit definition does not exhaust the possibilities of understanding notions and concepts. Indeed, one ought to admit the necessity to go one step further and assert that there is even a third kind of definitions as obtained by the *identification of opposites*. Yet while the two first kinds of definitions are current within mathematics and the sciences, the third kind is not encountered there precisely because mathematics and the ordinary sciences are modes of apprehension of finiteness and of multiplicity, whereas there are ideas in human understanding which allow to escape finiteness and multiplicity in order to apprehend that which outruns them. The two most formidable examples are on the whole *freedom* and above all *love.*

NICOLAS BERDIAEFF explained this in *The Reign of Spirit and the Reign of Caesar* by saying that Man is not only a finite being but also an infinite one and that in the latter is to be found the manifestation of the Divine. As far as we are concerned, we could even add that Man is not only both a finite and an infinite being, but that, being actually a *person*, he partakes into an exceptional unicity which contrasts against his belonging to the multiplicity of beings of World. That unicity as well as the infiniteness of which BERDIAEFF speaks make Man close to God.

Moreover, BERDIAEFF comments this by calling the attention upon the circumstance that we encounter here a specification of love, and he concludes that "a world without God (transcending it) would imply an insuperable contradiction between finitude and infinitude". This is indeed a proof of God, yet a proof based on the third kind of definition as well as on an experience of which he proposes a description under the form of a spiritual aspiration. Anyhow, BERDIAEFF thinks that the discourse which he develops is not part of any metaphysics proper, because it does not proceed from an objective attitude, but from an attitude which he holds for mystic, since it pertains to love. Yet this is not so certain and can be contended in the measure within which his discourse is very much objective and for the reason that love is not exclusively the motor of mystics, - whereas we must agree to say that the mystic experience of love is not of objective, but is of contemplative nature; i.e. it takes place 'within the temple *(cum templo),* in other words *'within the abode of God,'* where the (human) subject does not maintain the initiative of judgment upon his object, but transfers it to the Lord of the Place, accepting to be judged by Him. In such case, this is where the mystery of the divine be revealed to him, yet this does not make for him out of the enterprise a mysteriology, which would be a kind of contradiction in terms. All the same, in another passage, BERDIAEFF admits that "truth is infinite" i.e., in our understanding, that it must be possible to extend S. THOMAS' *adæquatio* to an adæquation between a discursive vision of the Capital Being and the experience which Man may have of Him.

For our part, love ought in no way to be relegated in, nor limited to, the domain of mystics. Love is ***the motor of all knowledge:*** objective, subjective, conjective and contemplative, since it is (alone) able to suspend temporality, and because it precedes all current principles of a so-called prime philosophy. (This has been explained at length elsewhere long ago in A. MERCIER, *De l'amour et de l'être,* Paris 1956, as well as again more recently in *Roseau pensant* (vol. I, II & III, Berne-New York-Paris 1988-92).)

Therefore, within love as the motor of all knowledge, one encounters the nucleus of the whole problem of the ***possibility of proofs of God.***

We could go so far as to say that Love, Death and Freedom 'in their togetherness' constitute the indubitable proof of God. The big question is: which is exactly that togetherness? This, however, is no more a question within the strict problematics of God, World, and Time.

(8) Nonetheless, for all that, one should not fall into the mistake to confuse love with a value.

We know from what has been in earlier sections aforesaid, that a value is to be regarded as that which is at the encounter of quality and quantity. LOUIS LAVELLE has by the way put forward exactly the same idea in his famous *Traité des valeurs* (1950).

Now, RENÉ LE SENNE, in his *Introduction à la philosophie* (1939), meant to establish a kind of identification between God and Value and proceeded by way of consequence to what he declares to be a proof of God.

If this holds, if we are right and if LE SENNE is not mistaken, God, in some understanding equal with Value (call it the Divine), must on the one hand be both Quality and Quantity at a time. In a profound sense, He is indeed, yet with the caution due to the incommensurability between human measures and the 'quantity' which would be His (which in religion is wittingly designated as 'omnipotence'). As regards His quality, it is pretty evident that it consists in the care which He takes about Time and World which He has detached and still detaches from Himself. Many a one will possibly object that such a care appears very neglected in view of the evils, the horrors and the lies encountered in World; yet such an objection rests on a grave misunderstanding, for we selves, the human, are those who produce evil, horror and lie in World, and not God Who, on His part, is engaged in effacing them along Time which comes and encounters us, in order to preserve World in its original perfection consisting in being the best of all possible worlds. This means that one more kind of argument is thus added to the former ones already encountered; LE SENNE calls it *the axiological argument.* Possibly this is the most stringent one!

On the other hand, if love were a value, as LE SENNE also (but wrongly) claims: which would be the quality and the quantity due to precisely make it a value? There are not such, for, instead of being a value, love is a 'motor', yet a motor which is under the rule neither of the first, nor of the second principle of thermodynamics, the reason why it is capable of doing, that the human soul travel as far as till the Absolute, Infinite and One, whose name is God and, consequently - without itself being a value - of promoting values up to the level of divine - not just human - perfection (what PASCAL used, in the booklet called *Sur la conversion du pêcheur,* to explain by saying that *"cette élévation ... ne peut (s')arrêter qu'elle ne soit rendue jusqu'au trône de Dieu"*). At the same time, one will notice that if one believes to love a particular being itself finite and a member of the multiplicity of beings belonging to World, one actually loves God through the very transparency of that being; however, the beings of and in World are not perfectly transparent, owing to the break of God's own Symmetry: they contain flaws like crystals having impurities, so it happens that one fails 'seeing' God through them, and the love shades off, the reason why one feels doubts about the vision being actually the vision of God, so much so that the 'proof of God' which our love was to supply, does escape us.

Finally, LE SENNE reminds us that beside those kinds of arguments, there is that of the *to-be-necessary-to-the-heart.* There again, we are reminded of PASCAL. This clearly needs no further comment.

After all, truth is, itself, a value just as the good, beauty, or even the divine or the sacred are. Many a contemporary does not share that opinion, but is wrong. S. AUGUSTINE had made of that his own argument, concluding that *"Truth ... must identify itself with Him Who Is"* (cf. *De immortalitate animæ,* XXII, 19). In such perspective, God would be 'proved' as being, if not *the* source, yet at least *at* the source of values, or of Value, as should be, an assertion which scandalized J.-P. SARTRE who saw in it a contradiction against the prejudice dear to existentialism claiming that freedom is nothing but 'man's capacity to create all by himself his own values'. Such existentialistic - typically atheistic - opinion has been criticized by H. DUMÉRY. We have contributed our share in its refutation. Apart from the loathsome arrogance it implies, as if Man were capable to remake all alone a world that should be better, more beautiful, more true than that of a - divine or spontaneous - creation (in which, - 'tis true, - evil, uggliness and deceitfulness appear to reign), existentialism has never been able to promote the creation of a better world. For, of course, either Man is part of a World which is radically self-sufficient and then any new arrangement which he introduces will never amount to the obtainment of an absolute perfection or anything else than a dissipation of energy substituting old orders with new ones, imperfect as well, or else there is a Source of Value or of increase, and then the question arises to know, what that source consists in and where it is to be found.

The axiological argument is, by the way, connected with the Cartesian proof based on the idea of perfection as summed up in the third *Meditation,* in the passage which runs as: *"je n'aurais pas l'idée d'une substance infinie, moi qui suis fini ...",* that is: imperfect, unless *"les perfections que j'attribue à ... Dieu ... (soient) en moi en puissance ..."* And that is after all what BERDIAEFF asserts when he says that Man is both finite and infinite at the same time.

(9) One could write pages explaining in detail the rôle assumed by love, by freedom, by death, in this 'Introduction to metaphysics conceived as the pursuit of the proof of God'. A reference to *death* has already been made in relation with the appearance and disappearance of entities in World. Let us simply add here that death, this time in conformity with an opinion held by many an existentialist (whether atheist

or believer), is to be considered the most important moment of life. Our personal reason in favour of that assertion is to be found in the circumstance that death is the testimony of *radical difference,* and that if there were not that difference, no idea could occur to any human mind. Death is therefore in principle ***at the origin of all ideas,*** in particular of the idea of God, and since at the same time it is the radical and ultimate possible ***experience,*** it grounds, if not the source of, yet at least the opening of the way to, metaphysics as theology, and in that sense ***death yields a proof of God***: Shall we call it the thanatological proof?

Love has also been dealt with in short, insisting that it alone puts us in a position to master the incommensurability between the finitude as combined with the multiplicity of our means of measure, and the infinitude as combined with the uniqueness of God. This assertion does not prove God under the wording as used presently, but it should be quite clear that without love, we could prove neither Him nor even anything about World... Yet the latter is quite another problem! (Cf. ***De l'amour et de l'être***, op. cit. further above).

Nothing precise has been said about freedom, except that the only definition we can give of authentic freedom consists in asserting that it dwells in the fact that we are not free, and that we have to accept it. That is an experiential fact, yet the definition is theoretical. That view on the nature of freedom has been commented upon and justified in several publications; it must be understood as the truth about freedom. It will suffice to recall that the most radical non-freedom we can be caught in resides in the inexorability of death. So that if we accept that we must and shall die, we are infinitely free. (The Reader may think for himself which is exactly the signification to be given to the adverb 'infinitely' just used.)

*

For all that, it is reasonable to claim that the meditation about, and thinking over, these three elements: Death, Freedom, and Love - *μελέτη περί θανάτου, ἐλευθερίας καί ἔρωτος* - constitutes the royal path of *metaphysics as theology;* and we might even say, inspite and against KANT: *as theodicy* i.e., as vindication of God. This is by the way not only in conformity with the Christian ideal, but so close to the philosophical complementary image one can win and realize out of Christianity, the very myth of which consists in a madness of love and whose meaning roots in the end of ends: ***in the eschaton,*** that the said metaphysics can right away be declared plain Christian metaphysics.

*

So finally, our conclusion will be the following: Either there is God, or there is no god: It has been explained further above that if a perfect God creates World, He cannot do it but at the cost of His perfection, since creation - by tearing World from His own Being - amounts to a break of His perfect symmetry, the consequence of which being the presence of imperfection i.e., of evil, horror ..., lie beside the Good, Beauty and Truth within World. Hence Man may very well fall a pray for evil, there being God or not. But if there is no god, who can pardon man for doing evil rather than the good? Either he be punished or he be pardoned. If there is no god, an impersonal justice cannot but punish him , - we remember that a person is one 'who can talk in the first person', - pardon as an alternative not being the business of justice, since the two are precisely complementary. But if, and only if, there is God, man may beg Him 'to be forgiven his offences just as he forgives those by whom he is offended' (cf. *Lord's Prayer*). Then philosophy and religion meet at a level which cannot be outdone. And this, we claim, might then very well be understood as a proof, even as the best among possible proofs of God.

- o -

Index

OF NAMES

- o -

Index

OF TERMS

[1]the word itself is used so frequently in the text, that only important mentions, yet devoid of particular qualifications, are quoted under that plain item

- o -

List of Publications by André Mercier

I BOOKS IN THE DOMAIN OF PHILOSOPHY (INCLUDING SUCH AS DEAL WITH SCIENCE IN GENERAL AND PHYSICS IN PARTICULAR)

1 *Logik und Erfahrung in der exakten Wissenschaft, unter besonderer Berücksichtigung der Physik* (Verlag A. Francke, Bern 1941, 62 pp.)

2 *Stabilité, complémentarité et déterminabilité* (F. Rouge éd., Lausanne 1942, 76 pp.)

3 *De la science à l'art et à la morale* (Ed. du Griffon, Neuchâtel 1950, 181 pp.)

4 *Thought and Being. An Investigation Into the Nature of Knowledge* (Verlag für Recht und Gesellschaft, Basel 1959, pp. I-X, 1-156)

5 *De l'amour et de l'être. Essai sur la connaissance* (Nauwelaerts, Louvain & Béatrice-Nauwelaerts, Paris 1960, pp. 1-193)

6 *Antikes und modernes Denken in Physik und Mathematik. Eine philosophische Betrachtung* (Band 13 der Schriftenreihe *Erfahrung und Denken,* Verlag Duncker & Humblot, Berlin 1964, pp. 1-154)

7 *Erkenntnis und Wirklichkeit* (Francke Verlag, Bern & München 1968, pp. 1-186)

8 *Science and Responsability. An Essay on the Theory of Values* (Studi internazionali di Filosofia, Torino 1969/70. pp. 1-126)

9 *Metaphysik, eine Wissenschaft sui generis. Theorie und Erfahrung auf dem Gebiet des Inkommensurablen* (Band 61 der Schriftenreihe *Erfahrung und Denken,* Verlag Duncker & Humblot, Berlin 1980, pp. 1-229)

10 *El tiempo, los tiempos, y la filosofía* (I. Ex. Wi., Berna 1985, pp. I-XII, 1-297)

11 *Roseau pensant* (Peter Lang éd., Berne-Francfort/M.-New York- Paris, Vol. I, avec une Préface par HENRI GOUHIER, pp. I-VIII, 1-218, 1988, Vol. II, pp. 1-362, 1989 et Vol. III, pp. 1-206. 1992)

II BOOKS IN THE DOMAIN OF PHYSICS

1 *Theoretisch-physikalische Fragen der Kosmologie* (Verlag Paul Haupt, Bern & Leipzig 1939, pp. 1-19)

2 *Leçons et problèmes sur la théorie des corps déformables* (F. Rouge & Cie, Lausanne 1943, pp. 1-156)

3 *Précis de physique générale. I: Introduction à la physique et à la mécanique* (Editions du Griffon, Neuchâtel 1945, pp. 1-197)

4 *Leçons et problèmes sur l'équilibre statistique et l'évolution de la matière* (Editions du Griffon, Neuchâtel 1945, pp. 1-186)

5 *Leçons sur les principes de l'électrodynamique classique* (Editions du Griffon, Neuchâtel 1952, p. 1-75)

6 *Principes de mécanique analytique* (Tome V du *Traité de Physique théorique et de physique mathématique* sous la direction de J.-L. DESTOUCHES, Gauthier-Villars éd., Paris 1955)

7 *Analytical and Canonical Formalism in Physics* (First edition, North Holland Publ. Co., Amsterdam 1959. Second edition, Dover Publications, Inc., New York 1963, pp. I-VIII, 1-222)

III BOOKS IN COLLABORATION

A. IN THE DOMAIN OF MATHEMATICS

1 ELIE CARTAN, *Leçons sur la théorie des spineurs* (2 volumes, Actualités scientifiques et industrielles No. 643, Hermann & Cie, éd., Paris 1938, rédigées d'après les notes recueillies par A. MERCIER)

B. IN THE DOMAIN OF PHYSICS

2 PAUL A. MERCIER et A. MERCIER. *La chaleur* (Tome II du *Précis de physique générale,* Editions du Griffon, Neuchâtel 1947, pp. 1-104)

3 A. MERCIER, H.-J. TREDER, W. YOURGRAU, *On General Relativity* (Akademie-Verlag, Berlin 1979, pp. 1-154)

C. IN THE DOMAIN OF PHILOSOPHY

4 F. RENOIRTE et A. MERCIER, *Philosophie der exakten Wissenschaften* (Bd. 8 der *Philosophia lovaniensis* in deutscher Ausgabe, Benziger Verlag, Einsiedeln, Zürich, Köln 1955). 2. Teil: *Mathematische Abstraktion und Naturbeschreibung*

5 ANDRÉ MERCIER UND JONATHAN SCHAER, *Die Idee einer einheit-*

lichen Theorie. Beitrag zur Methodologie der modernen Physik (Bd. 14 der Schriftenreihe *Erfahrung und Denken*, Verlag Duncker & Humblot, Berlin 1961, pp. 1-120)

D. IN THE SOCIAL DOMAIN

6 A. MERCIER und W. BAUMGARTNER, *Der Physiker* (Schriftenreihe *Die akademischen Berufe*, Nr. 5, Verlag Paul Haupt, Bern 1958)

IV FOUNDER-EDITOR (Presently Honorary Editor)

1 *General Relativity and Gravitation:* GRG Journal (Plenum Press, London & New York, Vol. I, 1970 *et ss.*)

V EDITIONS and CO-EDITIONS

1 *Fünfzig Jahre Relativitätstheorie - Cinquantenaire de la Théorie de la Relativité - Jubilee of Relativity Theory* (Verhandlungen - Actes - Proceedings, éd. par ANDRÉ MERCIER et MICHEL KERVAIRE, Helvetica Physica Acta, Suppl. IV, Birkhäuser Verlag. Basel 1956)

2 *Limites et critères de la connaissance* (Actes des Entretiens d'Oberhofen 1960 de l'Institut International de Philosophie, édités par ANDRÉ MERCIER avec l'assistance de M. P. LOCHER, Nos. 1 & 2 du Tome 15 de Dialectica, Paris et Neuchâtel 1961, pp. 1-336)

3 *Philosophes critiques d'eux-mêmes - Philosophers on Their Own Work - Philosophische Selbstbetrachtungen* (Série publiée sous les auspices de la Fédération Internationale des Sociétés de Philosophie FISP, éditée par ANDRÉ MERCIER et MAJA SVILAR, Verlag Herbert Lang, puis Peter Lang succ., Berne & Francfort/M., 1969 ss.)

4 *Mystik und Wissenschaftlichkeit* (Kulturhistorische Vorlesungen 1970/71 an der Universität Bern, hg. von ANDRÉ MERCIER, Verlag Herbert Lang, Bern & Frankfurt/M. 1972, pp. 1-190)

5 *Freiheit. Begriff und Bedeutung in Geschichte und Gegenwart* (Kulturhistorische Vorlesungen 1971/72 an der Universität Bern, hg. von ANDRÉ MERCIER, Verlag Herbert Lang, Bern & Frankfurt/M. 1973, pp. 1-194)

6 *Kunst heute. Betrachtungen im Lichte zeitgenössischer Tendenzen* (Kulturhistorische Vorlesungen 1972/73 an der Universität Bern, hg. von ANDRÉ MERCIER, Verlag Herbert Lang, Bern & Frankfurt/M. 1974, pp. 1-165)

[7] *Der Flüchtling in der Weltgeschichte. Ein ungelöstes Problem der Menschheit* (Kulturhistorische Vorlesungen 1973/74 an der Universität Bern, hg. von ANDRÉ MERCIER, Verlag Herbert Lang, Bern & Frankfurt/M. 1974, pp. 1-318)

[8] *Islam und Abendland. Geschichte und Gegenwart* (Kulturhistorische Vorlesungen 1974/75 an der Universität Bern, hg. von ANDRÉ MERCIER, Verlag Herbert Lang, Bern & Frankfurt/M. 1976, pp. 1-317)

[9] *Aggression und Toleranz. Wesen und Unwesen menschlichen Verhaltens* (Kulturhistorische Vorlesungen 1975/76 an der Universität Bern, hg. von ANDRÉ MERCIER, Verlag Herbert Lang, Bern & Frankfurt/M. 1977, pp. 1-210)

[10] *Umwelt und Mensch. Körperliche und seelische Auswirkungen* (Kulturhistorische Vorlesungen 1976/77 an der Universität Bern, hg. von ANDRÉ MERCIER, Verlag Herbert Lang, Bern & Frankfurt/M. 1978, pp. 1-124)

[11] *L'espace* (Actes des Entretiens de Berne 1978 de l'Institut International de Philosophie, édités par MAJA SVILAR et ANDRÉ MERCIER, Peter Lang éd., Berne & Francfort/M. 1987, pp. 1-273)

VI ARTICLES AND PAPERS IN THE DOMAIN OF GENERAL PHILOSOPHY, OF THE PHILOSOPHY OF ART AND OF METAPHYSICS

[1] *Sur les opérations de la composition musicale* (Archives de Psychologie, XXVII, 1939, p. 186ss.)

[2] *Classification des humanités* (Synthèse, VII, 1948, p. 154ss.)

[3] *Videnskaben og Ansvaret* (in Danish, Tidenskvinder, Aargang 1948, Nr. 41) [*Science and Responsibility*]

[4] *Esquisse d'une théorie unitaire des sciences et des arts* (Actes du XIe Congrès International de Philosophie, Bruxelles, Vol. X, North Holland Publ. Co., Amsterdam 1953, p. 207ss.)

[5] *Conscience du nécessaire et conscience héroïque* (Actes du VIIe Congrès des Sociétés de philosophie de Langue française: *La vie et la pensée,* PUF, Paris 1956, p. 168ss.)

[6] *Mise en valeur du vrai, du beau et du bien dans les relations humaines* (Actes du VIIIe Congrès des Sociétés de philosophie de Langue française, PUF, Paris 1956, p. 168ss.)

[7] *Le dilemme de la philosophie* (Theoria XXII, 1956, p. 168ss.)

[8] *Le paradoxe épistémologique et sa solution pratique* (Actes du IXe Congrès des Sociétés de philosophie de Langue française, *in* Etudes Philosophiques No 3. Aix-en-Provence 1957, p. 240ss.)

[9] *La connaissance considérée comme libération de la temporalité* (Anais del Congreso Internacional de Filosofia, São Paulo 1956, p. 217ss.)

[10] *Science et Philosophie* (Dialectica 11, 1957, p. 276ss.)

[11] *Wissenschaft und Philosophie* (Technische Rundschau, 49. Jg., Nr. 54, 1957, p. 3ss.)

[12] *Juger et ne pas juger* (Revue de théologie et de philosophie, tome VII, 1957, p. 206ss.)

[13] *Note sur l'interprétation de la causalité* (Synthèse, XI, 1959, p. 153ss.)

[14] *Das geistige Weltbild der Naturwissenschaft* (in *Das Weltbild der modernen Naturwissenschaft*, Hessische Hochschulwochen für staatswissenschaftliche Fortbildung 1952. Bd. 25/26, Verlag Gehlen, Bad Homburg-Berlin-Zürich 1960, pp. 168-188)

[15] *Tea for Two. A piece of fiction* (Dialectica 14, 1960, p. 121)

[16] *De la réalité et du mythe* (Dialectica 14, 1960, p. 215)

[17] *Science and Reality* (in *Traditional cultural values, East and West* (Entretiens of the International Institute of Philosophy and Indian Congress of Philosophy, Mysore 1959, pp. 43-49 and 125-127)

[18] *Limites et critères de la connaissance* (Actes des Entretiens d'Oberhofen 1960 de l'Institut International de Philosophie, Dialectica, Nos. 1 & 2, Tome 15, 1961: *Editorial,* pp. 9-11, *Limites et critères de la connaissance, Thème des Entretiens,* avec une discussion, pp. 25-38, 49-51, *Discours de clôture,* p. 328-329)

[19] *Des valeurs du vrai, du beau et du bien* (Les Etudes Philosophiques, 16e année, 1961, p. 407ss.)

[20] *L'évolution de la pensée au XXe siècle et le déclin de l'intuition sensible* (Revue Internationale de philosophie, 17e année, No. 64, 1963, p. 171ss.)

[21] *L'importanza filosofica dell'opera di Niels Bohr* (Filosofia, Separato No 22 di Filosofia delle Scienze, Torino 1964)

[22] *L'objectivité est-elle une condition de la vérité?* (Actes du 12e Congrès des Sociétés de philosophie de Langue française, Bruxelles-Louvain 1964, p. 118ss.)

[23] *Du droit, du devoir et de la liberté* (Actes des Entretiens de l'Aquila 1964 de l'Institut International de Philosophie sur *Le Fondement des Droits de l'Homme,* La Nuova Italia Editrice, Firenze 1966, p. 102ss.)

[24] *Pascal, l'homme, le nombre et l'infini* (Annales de la Faculté des Sciences de l'Université de Clermont, No. 7, 1er fascicule, 1962, p. 23ss.: Actes du Colloque de mathématiques, réuni à l'occasion du Tricentenaire de la mort de BLAISE PASCAL, T. I)

[25] *Ferdinand Gonseth et le problème du temps* (Studia philosophica, XXIV, 1964, p. 176ss.)

[26] *Voluntà di dialogo e promozione all'esistenza* (La Cultura, anno III, 1965, p. 374ss.)

[27] *Two Giants - Albert Einstein and Niels Bohr* (Technion, I, No. 2, 1965, p. 4ss.)

[28] *La Science, l'Etat et la Société* (Revue Suisse de la Relève Professionnelle et Scientifique, 6e année, Fasc. 1/2, 1967, p. 167ss.)

[29] *Fugit irreparabile tempus* (Berner Rektoratsrede, Verlag Paul Haupt, Bern 1967)

[30] *La Science au deuxième tiers du XXe siècle* (in *La Philosophie contemporaine. Chroniques,* éd. par R. KLIBANSKY, La Nuova Italia Editrice, Firenze 1968, vol. II, p.16ss.)

[31] *On the Foundation of Man's Rights and Duties* (Man and World, An International Philosophical Review, 1, 1968, p. 524ss.)

[32] *Knowledge and Physical Reality* (in *Physics, Logic and History,* Plenum Press, New York 1970, p. 39ss.)

[33] *La modalité du jugement comme notion clef d'une théorie générale de la connaissance* (in *La Modalité du jugement chez Aristote et dans la logique moderne,* Annale Universitatii Bucuresti, Seria Acta logica, 12, 1969, p. 153ss.)

[34] *Science and Responsibility* (in *Induction, Physics and Ethics,* Colloquium on the Philosophy of Science, Salzburg 1968, ed. by P. WEINGARTNER and G. ZECHA, D. Reidel Publ., Dordrecht 1970). *Idem* (Akten des XIV. Intern. Kongresses für Philosphie, Wien 1968, Bd. VI, Wien 1971, p. 412)

[35] *De l'évidence* (in *Problèmes de la théorie de la connaissance,* Entretiens d'Helsinki 1970 de l'Institut International de Philosophie, éd. par H. G. VON WRIGHT, Martinus Nijhoff, Den Haag 1972, p. 32ss.)

[36] *On Freedom* (Indian Philosophical Annual, 6, 1970, p. 36ss.)

[37] *Philosophie, Mensch und Freiheit* (Wissenschaft und Weltbild, 25. Jg., 1972, p. 83ss.)

[38] *Mystik und Vernunft* (in *Mystik und Wissenschaftlichkeit,* Kulturhist. Vorlesungen 1970/71 an der Universität Bern, hg. von ANDRÉ MERCIER, Verlag Herbert Lang, Bern & Frankfurt/M. 1972, p.355ss.)

[39] *Notes on the Analysis of Structure and Structuralist Ideologies. A Facet of Current French Philosophy* (Inquiry, 15, 1972, p. 355ss.)

[40] *L'oggettività della scienza è trasponible nell'arte e nella morale?* (Filosofia, XXIII, 1972, p. 375ss.)

[41] *Contemplation versus Reason. An Inquiry into Human Knowledge in the Light of a Theory of Value* (Religious Studies, 9, 1973, p. 49ss.)

[42] *Was bedeutet Freiheit?* (in *Freiheit. Begriff und Bedeutung in Geschichte und Gegenwart,* Kulturhistorische Vorlesungen 1971/72 an der Universität Bern. hg. von ANDRÉ MERCIER, Verlag Herbert Lang, Bern & Frankfurt/M. 1973, p. 9ss.)

[43] *Science et Responsabilité* (Relata Technica, 5, 1973, p. 389 & 6, 1974, p. 223ss.)

[44] *Propos sur la liberté* (in *Humanismo pluridimensional,* Atas da Primeira Semana Internacional de Filosofia, Soc. Brasileira de Fil. Cat., Ediçoes Loyola, São Paulo 1974, Vol. I, p. 43ss.)

[45] *Ethics and Science, and Responsibility in Technology* (Ibid., Vol.II, p. 586ss.)

[46] *Das Phänomen Kunst* (in *Kunst heute, Betrachtungen im Lichte zeitgenössischer Tendenzen,* Kulturhistorische Vorlesungen 1972/73 an der Universität Bern, hg. von ANDRÉ MERCIER, Verlag Herbert Lang, Bern & Frankfurt/M. 1974, p. 9ss.)

[47] *Science and Art* (in *Physical Reality and Mathematical Description,* ed. by CH. ENZ and M. MEHRA, D. Reidel Publ. Co., Dordrecht and Boston 1974, p. 3ss.)

[48] *Die Kunst als subjektive gegenüber der Wissenschaft als objektive Erhenntnis* (in Bregenzerwälder Kulturtage, Schriftenreihe, Vigil-Druck, Dornbirn 1974, p. 5ss.)

[49] *Science as Theory and as Praxis: A Main Subject-Matter of Contemporary Philosophy* (in *Philosophy: Theory and Practice,* Proceedings, Seminar on World Philosophy, Madras 1970. Center for Advanced Study

in Philosophy, University of Madras 1974, p. 494ss.)

[50] *Philosophy is not Theory (Comments)* (Ibid., p. 33)

[51] *Prospects of Anthropological Rationalism (Comments)* (Ibid., p.76)

[52] *Theory and Practice in the Evolution of Western Thought (Comments)* (Ibid., p.288)

[53] *Science and Religion, and Their Complementary Nature* (Special Lecture at the World Philosophy Conference, Golden Jubilee Session of the Indian Philosophical Congress Dec. 1975/Jan.1976. In: *Knowledge, Culture and Value,* ed. by R. C. PANDEYA & S. R. BHATT, Motilal Banarsidass, Delhi-Varanasi-Patna 1976, Vol. I, p. 277ss.)

[54] *La philosophie comme rencontre dialectique des sciences, des arts, de la morale et de la mystique* (in *Philosophes critiques d'eux-mêmes.* publ. par ANDRÉ MERCIER et MAJA SVILAR, Verlag Herbert Lang, Bern & Frankfurt/M. 1976, Vol. 2, p. 149ss.)

[55] *Does Science Coincide with our Knowledge about Nature?* (Dialectics and Humanism, No. 1, 1976, p. 43ss.)

[56] *On Peace* (in *Philosophy East and West.* dedicated to Professor T. M. P. MAHADEVAN, ed. by H. D. LEWIS, Blackie and Son. Madras 1976, p. 208ss.)

[57] *Philosophie als Weg zur Toleranz* (in *Aggression und Toleranz, Wesen und Unwesen menschlichen Verhaltens,* Kulturhistorische Vorlesungen 1975/76 an der Universität Bern, hg. von ANDRÉ MERCIER, Verlag Herbert Lang, Bern & Frankfurt/M.. 1977, p.169ss.)

[58] *Qualità e Quantità sono incompatibili?* (in *La Qualità,* Seminari interdisciplinari di Venezia, a cura di E. R. LORCH, Società editrice Il Mullino, Bologna 1976, p. 401ss.)

[59] *Qualité et Quantité, ou l'Etre et l'Avoir* (in *Filosofia e realidade brasileira,* Atas II Semana Internacional de Filosofia, Petropolis 1974, a cura S.B.F.C., Compania editora Americana, Rio de Janeiro 1976, Vol. I, p. 18ss.)

[60] *O umetnosti i nauci* [*On the Arts and the Sciences*] (Književna Kritika Broj, 3, 1977, p. 25ss., in Serbo-Croat)

[61] *From Mechanism to Humanism and Further: For a Demythification of Determinism. Or: The Plurality of Determinisms and the Shortcoming of Humanism* (in *Christian Theology in the Context of Scientific Revolution,* Communications of the New York Symposium, July 1977, Académie Internationale des Sciences Religieuses et Académie Interna-

tionale de Philosophie des Sciences, ed. by M. FOUYAS, Reprinted from Abba Salama, Vol. IX, Athens 1978, p. 41ss.)

62 *Recherche d'une définition de la technique* (Revue philosophique du Maroc, 1978, p. 23ss.)

63 *Idem* (A Filosofia e as Ciências, Atas IV Semana internacional de Filosofia, Rio de Janeiro 1979, p. 11ss.)

64 *Philosophy and Civilization: Conclusions* (in *Philosophy and Civilization*, Proc. of the First Afro-Asian Philosophy Conference, 13-16 March 1978 in Cairo, ed. by MOURAD WAHBA, Cairo, Faculty of Education, Ain Shams University, Cairo 1978, p. 249ss.)

65 *De l'art et de la science: Métalégomènes d'une théorie axiologique de la connaissance* (Proceedings of the First International Week on the Philosophy of Art, Cephalonia 1976, Part II, Diotima, Vol. 6, 1978, p. 43ss.)

66 *La poésie comme métaphysique* (Diotima, Vol. 7, 1979, p. 208)

67 *Does Science Coincide with Our Knowledge about Nature?* (in *Man and Nature,* Second International Conference of the International Society for Metaphysics, Visva-Bharati, Santiniketan 1976. Ed. by G. F. MCLEAN, Oxford Univ. Press, Calcutta-Delhi-Bombay-Madras 1978, p. 15ss.)

68 *La menace des idéologies et du mysticisme en face de la véritable métaphysique* (Memorias del Symposium filosófico internacional en homenage al XXV aniversario de la Sociedad mexicana de filosofía, Ixtapan de la Sal, Estado de México, 5-11 Julio de 1979, Vol. I, p. 11ss.)

69 *Idem* (*Φιλοσοφία*, 8-9. 1978/79, p. 16ss.)

70 *Hat uns der Fortschritt weitergebracht?* (in *Was ist Fortschritt?* Kulturhistorische Vorlesungen 1978/79 an der Universität Bern, hg. von MAJA SVILAR, Verlag Peter Lang, Bern & Frankfurt/M., 1979, p. 183ss.)

71 *Technique et responsabilité du point de vue de la pratique philosophique* (in *Science, Technologie et Développement Africain,* A.A.A.S.T., Dakar 1978, p. 201ss.)

72 *Introductory Remarks,* and *Albert Einstein and Humanism - After-dinner Speech* (in: Albert Einstein Centenary Celebration in Berne 1979, Epistemologia III, 1980, No speziale, pp. 9-11 & 227-230)

73 *Metafísica del tiempo* (in *La Filosofía del Cristiano, Hoy* (Autos del Primer Congreso Mundial de Filosofía Cristiana, Embalse, Argentina

1979, ed. Sociedad Católica Argentina de Filosofia, Córdoba (Arg.). 1981, Vol. III, p. 1469)

[74] *Le viol des symétries comme principe structurel de la création artistique* (Actes de la Deuxième Semaine Internationale de Philosophie de l'Art, Samos 1979, Deuxième Partie, Diotima, Vol. 10, 1982, p. 72ss.)

[75] *Foundation of Human Rights,* and *Speech* (in *The philosophical Foundation of Human Rights,* Papers presented to the International Seminar, 9-13 June 1980 at Hacetepe University, Ankara, ed. by JOHANNA KUÇURADI, Meteksan Ltd., Ankara 1982, pp. 15-22, 185-187)

[76] *El tiempo desde el punto de vista de un físico "metafísico"* (Escritos de filosofía, Academia Nacional de Ciencias, Buenos Aires, 5: *Tiempo-1,* Año VII, Enero-Junio 1980, p. 878ss.)

[77] *Future of Religion in a Unified Civilization* (Proceedings of the First Islamic Philosophy Conference 19-22 Nov. 1979, Cairo on *Islam and Civilization,* ed. by MOURAD WAHBA, Ain Shams University Press, Cairo 1982, p. 117ss.)

[78] *Tolerance as a Philosophical Imperative* (Proceedings, First EASRG Regional Conference 21-24 Nov. 1981, Cairo on *Cultural Tolerance,* ed. by MOURAD WAHBA, The Anglo-Egyptian Book Shop, Cairo 1982, p. 53ss.)

[79] *Complémentarité des droits et des devoirs de l'homme* (Memoria del X Congreso Mundial Ordinario de Filosofía del Derecho y Filosofía Social, I.V.R., Vol. VIII, Universidad Nacional Autónoma de México. Instituto de Investigaciones Jurídicas, Serie G: Estudios doctrinales Num 70, México 1982, p. 27ss.)

[80] *A Physicist's View on the Mind-Body Problem* (Epistemologia, Vol. 4, Special Issue, 1981, p. 21ss.)

[81] *The metaphysical, the mystical and the religious Approach to Life* (in *Philosophy and Cultures,* ed. by H. ODERA ORUKA & D. A. MASOLO, Bookwise Ltd., Nairobi 1983, pp. 75-85: Proc. of the 2nd Afro-Asian Philosophy Conference in Nairobi, 1981)

[82] *Ueber die Vollkomenheit* (in *Wahrheit und Wirklichkeit,* Festgabe für LEO GABRIEL zum 80. Geburtstag, hg. v. P. KAMPITZ, G. PÖLTNER & H. VETTER, Duncker und Humblot, Berlin 1983, pp. 69-74)

[83] *Tecnica e tecnologia fra natura e cultura* (Nuova Civiltà delle Macchine, Anno II, n⁰ 2, Primavera 1984, pp. 19-25)

[84] Articles:
AGAZZI - AVINERI - BLANSHARD - BOCHENSKI - CALOGERO - CAWS - DAMNJANOVIC - DEGEORGE - DIEMER - EWING - FARBER - GABRIEL - GIACON - GRASSI - HARTSHORNE - HERSCH - HUEBNER - KOSING - LOMBARDI - MCKEON - MCLEAN - MAHADEVAN - MARKOVIC - NASR - PARAIN-VIAL - PARSONS - PASSMORE - PELC - PUCELLE - RADNITZKY - V. RINTELEN - SCHAFF - SMITH - THEODORACOPOULOS - TOPITSCH - WEISS in: *Dictionnaire des Philosophes,* ed. by D. HUISMAN for PUF, 2 vol., pp. I-XXIV, 1-2725, Paris 1984)

[85] *Epilogue* (in *Philosophie et Technique - Philosophy and Technology,* ed. by A. MERCIER, Entretiens de l'Institut International de Philosophie à Montreal 1983, Berne & Paris 1984, pp. 76-86)

[86] *L'intersubjectivité comme condition d'une compréhension transculturelle* (in *Pela filosofia, Pour la philosophie,* Homenagem a TARCISIO MEIRELES PADILHA, Mélanges, Pallas SA, Rio de Janeiro 1984, pp. 309-320)

[87] *Nauka a sztuka* (in *Estetyka w świecie,* red. MARIA GOLASZEWSKA, Universytet Jagielloński, Skrypty Uczelniane Nr. 485, Kraków 1984, pp. 163-182)

[88] Four Chronicles on *metaphysics* in Spanish in the 'Dominical' (Sunday issue), suppl. to El Comercio, daily newspaper of Lima (Peru): 1. *La Metafísica: teoría y experiencia del inconmensurable* (No 46, 11/10/1984, p.7). 2. *Sócrates, Platón, Aristóteles, Plotino y la metafísica* (No. 08, 24/2/1985, p. 8). 3. *Los orígenes de la metafísica* (No 23, 9/6/1985, p. 11 & 19). 4. *La metafísica, el conocimiento y los valores* (No. 35, 1/9, 1985, p.6)

[89] *Ne pas se contenter d'une illusion de l'infini* (in *L'Eidos de l'art,* éd. par le Séminaire d'esthétique de l'Institut de Philosophie de l'Université Jagellone à Cracovie, Colloque international d'Esthétique, Cracovie 23-29 août 1985)

[90] *Sztuka i Odpowiedzialnósi* (in *Ethos Sztuki, Materialy z Ogelnopolskiej, Eseji Estetycznej, Mogilany,* maj 1983. Praca zbiorowa pod redakcja MARII GOLASZEWSKIEJ, Panstwowe Wydawnictwo N., Warszawa-Krakow 1985, pp. 113-134)

[91] *Le carnaval considéré comme technique de vie* (in *Le Carnaval, la fête et la communication,* Actes des Premières Rencontres Internationales, Nice 8-10 mars 1985, Ed. Serre, Nice 1985, pp. 435-440)

[92] *Il tempo in fisica e metafisica* (Synesis, Anno II, n. 2/3, 1985, pp. 31-48)

93 *La rationalité de la science peut-elle - ou doit-elle - être soit dépassée, soit complétée?* (Epistemologia IX, 1986, 31-42)

94 *La dimension cosmique de l'œuvre d'art* (Diotima 14, 1986, pp. 28-40)

95 *L'à-venir, où se cache le dieu* (in *L'Avenir,* Actes du XXIe Congrès de l'Assoc. des Sociétés de Philosophie de Langue française, Athènes 1986, Librairie philosophique J. Vrin, Paris 1987, Conférences, pp. 38-54)

96 *Discours inaugural* (prononcé à l'Ouverture de l'Académie Internationale de Philosophie de l'Art le 23 septembre 1984 à Corfou, Actes de l'AIPA, Fasc. 1, 1986, pp. 7-11)

97 *Art, Science, Universal Values: The Quality of Contemporary Life* (Presidential Address for the Meeting of AIPA in New Delhi, Dec. 1985, in Actes de l'AIPA, Fasc. 3, 1987, pp. 11-17)

98 *Time and its Valuation in Art, Science and Metaphysics* (Actes de l'AIPA, Fasc. 3, 1987, pp. 155-170)

99 *Naissance et mort du langage* (dédié à ALAN HELD, Nouvel An 1988. Brochure sans lieu d'impression)

100 *Le vrai et le beau: unité de l'épistémologie, de l'axiologie et de l'ontologie* (Proc. 3rd Intern. Week on the Philosophy of Art, Corfu 1984, Part II, Diotima 15, 1987, pp. 73-77)

101 *La philosophie ne sert à rien, mais on ne peut s'en passer* (©A.M., I.T.P., Univ. de Berne 1988, pp. 1-21)

102 *L'expression "d'humanisme chrétien" est-elle "contradictio in adjecto"?* (in *El humanismo y la metafísica cristiana en la actualidad,* Secundo Congreso Mundial de Filosofía Cristiana, Monterrey, N.L., México, 1986, Vol. 1, no. 2, 1988, pp. 25-39)

103 *The Nature of Values and their Encounter - Their Relationship to Culture* (The Asian Journal of Philosophy, Vol. 1, no. 2, 1988, pp. 25-39)

104 *Chronos et Kairos* - Discours de synthèse (in *Chronos et Kairos,* Entretiens d'Athènes 1986 de l'Institut International de Philosophie, Diotima 16, 1988, pp. 66-82)

105 *Discours présidentiel,* Réunion de Madrid de l'AIPA, 4-7 septembre 1987 (Actes de l'Acad. Intern. de Philosophie de l'Art AIPA, Fasc. 4, 1988, pp. 1-19)

106 (avec M. SVILAR) *Toleranz als dialektische Methode - Gedanken zur Rolle der Philosophie heute* (in *Die Philosophie in der modernen Welt,*

Gedenkschrift für Alwin Diemer, hg. v. Ulrike Hinke-Doernemann, Verlag Peter Lang, Frkft./M.-Bern-New York-Paris, Teil I, pp. 79-90)

107 *La Paix, le Dieu, le Monde et le Temps (Śānti-Dharma-Éσχατον)* (in *Philosophia pacis, Homenaje a Raimón Panikkar,* Miquel Siguan ed., Simbolo editorial, Madrid 1989, pp. 17-35)

108 *Die Zu-Kunft, wo der Gott sich verbirgt* (Glaube und Denken, 2. Jg., pp.25-48, 1989)

109 *Le temps, le monde, et la liberté, Fondement de la métaphysique comme science sui generis de l'incommensurable* (Epistemologia XI, 1988: *Science et Métaphysique*, pp. 183-196)

110 *Le temps de la préhistoire et la venue de l'homme et des cultures* (*Φιλοσοφία*, Acad. of Athens, 1987-1988, pp. 153-167)

111 *Plea for a Philosophy of Literature and Poetry.* A letter to Mrs Eleanor Wikborg, Professor of English Literature at Stockholm University (Univ. of Berne, 1990, 13 p.)

112 *Νοεῖν καί Διαφερεῖν. La Réalité ultime et ses approches distinctes* (in *Du Vrai, du Beau, du Bien, études philosophiques présentées à Evanghelos Moutsopoulos,* Librairie Philosophique J. Vrin, Paris 1990: Collegium Academicum Universale Philosophiae, Publications Série " Recherches", N^0 1, pp. 69-90)

113 *La metafisica e i suoi legami con la scienza* (in *La probabilità nelle scienze,* Convegno di Vico Equense-Napoli, maggio 1987, a cura di Giuseppe del Re, Quaderni dell I.P.E., Napoli 1990, pp.67-74)

114 *La notion d'interprétation: Qu'est-ce qu'interpréter?* (Introd. au Thème de la Réunion de l'AIPA à Calamata, Grèce, en 1989 - in Diotima, 18, pp. 7-22, 1990, & Actes de l'AIPA, Fasc. 5, 1990)

115 *Odpovědnost, teorie hodnot a teorie porozumění* (Filosofický Časopis, Obsah Čisla 1, Ročník 1, 1991/39, Čekoslov. Akademia Věd. Praha, pp. 38-53)

116 *Poiésis, source et moteur de connaissance* (in *Poïétique,* Actes du 1er Colloque de philosophie de la création, du 28 avril au 2 mai 1989 à Vinneuf (F-89), pp. 219-224, 1991)

117 *Ecrire sur la mystique?* (in *Probleme philosophischer Mystik,* Festschrift für Karl Albert, hg. v. Elenor Jain & Reinh. Margreiter, Akademie-Verlag Skt Augustin 1991, pp. 15-20)

118 *The Problem of the Imperfection of a World, Itself created by a Perfect God,* dedicated to Sir Karl Popper on the occasion of his 90th

birthday (Foundations of Physics, Vol. 22 No.2, 1992, p. 205-219)

119 *El Problema de las Pruebas de Dios Hoy* (Memorias del Tercer Congreso Mundial de Filosofia Cristiana, Quito 1989,Vol. 2, pp. 213-230, Ed. de la Pontificia Univ. Cat. del Ecuador, Quito 1990)

120 *On Intersubjectivity, Friendship and Godliness* (in *Perspectives in Philosophy, Religion, and Art,* Essays in Honour of MARGARET CHATTERJEE, Indian Council of Philosophical Research, New Delhi 1993, pp. 20-33)

121 *Is there a "Philosophy of Art"?* (in *Primum Philosophari, Księga pamiątkowa Stefanowi Morawskiemu ofiarowana,* pod redakcjq J. BRACH-CZAINY, Oficyna Naukowa, Warszawa 1993, pp. 275-284)

122 *Some thoughts on Jaspers and Heidegger, Their Distance and Relationship* (in *Karl Jaspers, Philosopher among Philosophers,* ed. by RICH. WISSER & LEONARD H. EHRLICH, Königshausen & Neumann 1993, pp. 109-112)

VII ARTICLES AND PAPERS IN THE DOMAIN OF THE PHILOSOPHY OF THE PHYSICO-MATHEMATICAL SCIENCES

1 *Sur l'axiomatique de la théorie cinématique de Milne* (Helvetica Physica Acta, 13, 1940, p. 473ss.)

2 *Die Axiomatik der Quantentheorie und die Begriffsgestaltung in der Physik des Mikrokosmos* (Mitt. Naturforschende Gesellschaft, Bern 1942, p. 23ss.)

3 *Considérations épistémologiques sur la physique moderne* (Annuaire de la Société Suisse de Philosophie, Tome II, 1942, p. 141ss.)

4 *Constantes universelles* (Actes de la Société Helvétique des Sciences Naturelles, 1944, p. 61ss.)

5 *Lois de la Nature et Constructions de l'esprit* (Synthèse, V, 1946, p. 203ss.)

6 *Unité et Méthode dans la science* (Synthèse, V, 1946, p. 287ss.)

7 *Physique géométrique et physique de l'interaction, deux tendances* (Proc. Xth International Congress of Philosophy, Amsterdam 1948, p. 41ss.)

8 *Position philosophique de la physique parmi les sciences* (Gymnasium Helveticum 3, 1949, p. 187ss.)

9 *Les conditions physiques et la notion de temps* (Studia philosophica X, 1950, p. 85ss.)

10 *La notion d'irréversibilité* (Mitt. Naturforsch. Gesellschaft Bern, Neue Folge, 8. Bd., 1951)

11 ***Die Zeit und die Relativität der Kategorien im Lichte der modernen Physik - Il tempo e la relatività delle categorie nella fisica moderna*** (Archivio di Filosofia, *Il Tempo,* Anno 1959, No 1, pp. 191-207, 209-222)

12 ***Physique et Philosophie en Allemagne, en Autriche et en Suisse*** (in *La Philosophie au milieu du XXe siècle,* Chroniques de l'Institut International de Philosophie, éd. par R. KLIBANSKI, Vol. I, Florence 1958, p. 292ss.)

13 *La physique théorique* (Revue Universitaire Suisse, 35, 1962, p. 83ss.)

14 *Sur les relations possibles de la physique avec la philosophie contemporaine* (Helvetica Physica Acta, 36, 1963, p. 937ss.)

15 *Petits prolégomènes à une étude sur le temps* (Theoria, XXIX, 1063, p.277ss.)

16 *La physique et l'information* (in *Information and Prediction in Science,* Proceedings, Symposium of the International Academy for the Philosophy of Sciences, Brussels 1962, ed. by S. DOCKX and P. BERNAYS, Academic Press, New York and London 1965, p. 181ss.)

17 *On the Unification of Physical Theories* (Symposia on Theoretical Physics and Mathematics, ed. by ALLADI RAMAKRISHNAN, Plenum Press, New York 1968, Vol. 8, p.15ss.)

18 (With G. RÜTTIMANN) *Le problème posé à la logique par l'apparition des théories quantiques* (Les Etudes Philosophiques, 4, 1969, p. 461ss.)

19 *Thoughts on the Dynamics of Foundation,* or *What I believe* (Foundations of Physics, 1, 1971, p. 285ss.)

20 *Forms of Determinisms, Objectivity, and the Classification of Sciences* (in *Perspectives in Quantum Theory, Essays in honor of Alfred Landé,* ed. by W. YOURGRAU and A. v. D. MERWE, MIT Press, Cambridge Mass. 1971, & 2nd ed., Dover Publications, Inc., New York 1979, p. 244ss.)

21 *Epistemological Questions Concerning Cosmology and Gravitation* (in *The Nuclei of Galaxies, Black Holes and Collapsed Matter,* International School of Cosmology and Gravitation, IIIrd Course, Erice 1974)

22 *Idem* (GRG-Journal, Vol. 6, No 5, 1975, p. 513ss.)

[23] *Speculative Remarks on Physics in general and Relativity in particular* (in *Philosophical Problems in Physics,* Dialectics and Humanism, No 2, 1975, p. 125ss.)

[24] (reproduced in *Topics in Theoretical and Experimental Gravitation Physics,* ed. by V. DE SABBATA and J. WEBER, NATO Adv. Study Institutes Series, B. Vol. 27, Plenum Publ. Co., New York 1977)

[25] *Ferdinand Gonseth (1880-1975): Mathematik und Wirklichkeit. Ein Stück Geschichte der Wissenschaftsphilosophie in der Schweiz* (ZS für allg. Wissenschaftstheorie, VII/1, 1976, p. 11ss.)

[26] *Conjectures* (GRG-Journal, Vol. 8, 1977, p. 673ss.)

[27] *Birth and Role of the GRG-Organization and the Cultivation of International Relations among Scientists in the Field* (in *Albert Einstein, His Influence on Physics, Philosophy and Politics,* ed. by P. C. AICHELBURG and R. U. SEXL, Friedr. Vieweg & Sohn, Braunschweig-Wiesbaden 1979, p. 185ss.)

[28] *On Time, Space-Time and Other 'Time'* (in *Perspectives in Metascience,* dedicated to Professor HÅKAN TÖRNEBOHM, ed. by JAN BÄRMARK, Interdisciplinaria 2, Acta Regiæ Soc. Scientiarum et Litterarum Gothoburgensis, Göteborg 1979, p. 139ss.)

[29] *F. Gonseth, les Mathématiques et la Réalité* (Epistemologia, 4, no 1, 1981, p. 235ss.)

[30] *Physics and Philosophy* (in *Old and New Questions in Physics, Cosmology, Philosophy and Theoretical Biology,* in Honor of W. YOURGRAU, ed. by A. V. D. MERWE, Plenum Publ. Corp., New York 1983, pp. 671-681)

[31] *A propos de la classification des sciences* (in *Saggi di Storia del Pensiero Scientifico* ded. a VALERIO TONINI, a cura di V. CAPPELLETTI ed al., Jouvence, Roma 1983, pp. 185-199)

[32] *Considérations finales* (in *La Science face aux attentes de l'homme contemporain,* Colloque de l'Académie Internationale de Philosophie des Sciences 5-8 avril 1983 à Séville-La Rabida, Office Intern. de Librairie, Bruxelles 1984, pp. 294-296)

[33] *De la Vérité* (in *La nature de la vérité scientifique,* Archives de l'Institut International des Sciences Théoriques N° 27: Colloque Jubilaire à l'occasion des 80 ans du R.P. STAN. DOCKX, O.P.; CIAO, Louvain-la-Neuve 1985, p. 35-53)

[34] *François Viète, René Descartes et René-Anthoine Ferchault de Réau-*

mur, pionniers de la Science moderne dite exacte (Ann. de la Soc. d'Emulation de la Vendée, 1986, pp. 55-61)

35 *Differenti forme di determinismo* (Scuola e Cultura, Nuova Secondaria 1, 15 sett. 1986, pp. 37-39)

36 *Science as Theory and Practice* (in *Lectures on Philosophy and Physics,* ed. by MOURAD WAHBA. Cairo 1986, Fac. of Education, Ain Shams University, pp. 1-12)

37 *Time, the concept of gravitation and the possibility of a unified field theory* (ibid., pp. 13-21)

38 *Réflexions sur la science et son histoire* (in *Les relations mutuelles entre la philosophie des sciences et l'histoire des sciences,* Congrès de l'Académie Internationale de Philosophie des Sciences à Fribourg, Discours de clôture. Epistemologia X, 1987, pp. 249-265)

39 *Foreword* (to the Proceedings of the Meeting of the Internat. Academy for the Philosophy of Science in Cesena (I), May 1988) on *Science et Métaphysique,* in Epistemologia XI, 1988, pp. 3-10)

40 *General Relativity at the Turning Point of its Renewal* (in *Studies in the History of General Relativity,* Vol. III, p. 109ss. Birkhäuser eds., Boston-Basel-Berlin 1992)

VIII ARTICLES & PAPERS IN THE DOMAINS OF MATHEMATICS, PHYSICS AND SCIENCE PROPER

1 *Contribution à l'étude des dérives continentales* (Archives des Sciences physiques et naturelles, 15, fasc. 3/4, 1933)

2 (With H. SAINI) *Dilatation thermique du nitrate de sodium mesurée aux rayons X* (Helv. Physica Acta, VII, 1934, p. 267ss.)

3 *Application des nombres de Clifford à l'établissement du théorème de relativité de Lorentz* (Helv. Phys. Acta, VII, 1934, p. 649ss.)

4 *Relations entre la répartition des densités de la croûte terrestre et les valeurs de la pesanteur* (C. R. Soc. de physique et d'histoire naturelle, Genève, 51, 1934, p. 195)

5 *Idem* (Archives des Sc. phys. et naturelles, 16, 1934, p. 144ss.)

6 *Expression du second principe de la thermodynamique relativiste au moyen des nombres de Clifford* (C.R. Soc. de phys. et d'histoire naturelle, Genève, 52, 1935, p. 113)

[7] *Expression des équations de l'électromagnétisme au moyen des nombres de Clifford* (Thèse de Doctorat, Genève 1935)

[8] *Sur les nombres de Clifford: 1. Une formule d'intégration partielle. 2. Propriété d'une équation différentielle* (C.R. Acad. des Sciences, Paris, 201, 1936, p. 1320)

[9] *Effet d'une erreur d'estimation des densités du sial et du sima dans l'évaluation des anomalies de la pesanteur* (Archives des Sciences, 18, 1936, p. 21ss.)

[10] *Relation entre les grandeurs spinorielles et cliffordiennes* (C.R. Soc. phys. et d'histoire naturelle, Genève, 53, 1936, p. 102)

[11] *Sur la théorie de la radioactivité β* (C.R. Acad. des Sciences, Paris, 204, 1937, p. 1117)

[12] *A note on the theory of β-radioactivity* (Nature, 139, 1937, p. 797)

[13] *Sur l'équation* $\nabla \to C = \alpha C$ (C.R. Acad. des Sciences, Paris, 204, 1937, p. 797)

[14] *Matrices pouvant servir d'unités pour les nombres de Clifford* (Actes Soc. Helv. Sciences naturelles, p. 9l, 1937)

[15] *The liquid state of the primitive earth* (Nature, 141, 1938, p. 201)

[16] *Sur la liquéfaction du globe terrestre dans l'hypothèse d'une sphère gazeuse initiale soumise à une loi polytropique* (Archives des Sciences, 20, 1938, p. 31ss.)

[17] (With TORSTEN GUSTAFSON) *Remarque sur l'énergie propre de l'électron* (C.R. Acad. des Sciences, Paris, 206, 1938, p. 1217)

[18] *Sur l'énergie propre de l'électron* (Helv. Physica Acta, 12, 1939, p.55ss.)

[19] *Beziehungen zwischen den Clifford'schen Zahlen und den Spinoren* (Helv. Physica Acta, 14, 1941, p. 565ss.)

[20] *Précisions sur la liquéfaction du globe terrestre* (C.R. Soc. Phys. et d'Histoire naturelle, Genève, 59, 1942, p. 83)

[21] *Universelle Konstanten, Elementarteilchen und fundamentale physikalische Beziehungen, I, II und III* (ZS für den physikalischen und chem. Unterricht, 55, 1942)

[22] *Remarque sur la dynamique des fluides visqueux* (Helv. Physica Acta, 15, 1942, p. 509)

[23] *Sur le champ du tourbillon dans les fluides* (C.R. Soc. de phys. et d'histoire naturelle, Genève, 59, 1942, p. 239)

[24] *Constantes universelles* (C.R. Soc. de phys. et d'histoire naturelle, Genève, 60, 1943, p. 214)

[25] *Constantes universelles, unités naturelles et invariance relativiste* (C.R. Soc. de phys. et d'hist. naturelle, Genève, 60, 1943, p. 256)

[26] *Analyse dimensionnelle et analyse géométrique ou algébrique; les caractères d'une grandeur physique* (Helv. Phys. Acta, 17, 1944, p. 242)

[27] *Constantes universelles* (Actes de la Soc. Helv. des Sciences Naturelles, 1944, p. 61ss.)

[28] Hypothèse d'identité entre le fin et le grossier comme base de la statistique (Experientia, I/3, 1945)

[29] (With ED. KEBERLE) *L'énergie et le temps, et les relations canoniques* (Archives des Sciences, 2, 1949, p. 186)

[30] *Sur le formalisme canonique, la condition accessoire $H + p_{f+1} = 0$, et la réversibilité de la mécanique classique et de la théorie des quanta* (Archives des Sciences, 2, 1949, p. 403)

[31] (With ARNOLD WERFELI) *Le théorème d'adiabatie dans le formalisme canonique homogène* (Archives des Sciences, 2, 1949, p. 580)

[32] *Sur les fondements de l'électroodynamique classique* (Archives des Sciences, 2, 1949, p. 584)

[33] *Relativité et statistique* (Proc. Kon. Nederl. Akad. van Wetensch., LIII, 1950, p. 50)

[34] (With ED. KEBERLE) *Comportement relativiste des systèmes avec ou sans self-contrainte* (Archives des Sciences, 3, 1950, p. 235)

[35] *Objection possible à une théorie de la lumière* (Helv. Phys. Acta, 23, 1950, p. 50)

[36] *Relativité et statistique* (Congr. internat. de Phil. des Sciences, Paris 1949, Actes, Paris 1951, p. 65)

[37] *Die Idee der Quantentheorie des Wellenfelder* (Physikalische Blätter, 8, 1952, p. 440)

[38] (With P. WILKER) *Remarques sur la singularité du temps, l'utilisation d'un formalisme quantique homogène et sur la relation d'incertitude entre le temps et "l'énergie"* (Helv. Phys. Acta, 26, 1953, p. 181)

[39] (With M. KERVAIRE) *Albert Einstein und die Lehren der Mechanik und der Elektrizität* (Technische Rundschau, Jahrgang 47, Nr. 20 & 21)

[40] *Fifty Years of the Theory of Relativity* (Nature, 175, 1955, p. 919)

[41] (With P. Wilker) *Homogener kanonischer Formalismus und Quantentheorie* (Proc. 8th Intern. Congress of Pure and Applied Mechanics, Istanbul, 1955, p. 519)

[42] (With Edith Schaffhauser) *Théorie unitaire des champs gravitationnel et électromagnétique* (Proc. 8th Intern. Congress of Pure and Applied Mechanics, Istanbul, 1955, p. 521)

[43] (With V. Gorgé et W. Lindt) *Programme d'analyse des hyperfragments pour la Bull-gamma AET* (Archives des Sciences 13, 1960, p. 139)

[44] *Report on the Dallas Conference on Gravitational Collapse and Other Topics in Relativistic Astrophysics* (Bull. GRG, No 6, 1964, p. 1)

[45] *Report on the Second Texas Symposium on Relativistic Astrophysics* (Bull. GRG, No 8, 1965, p. 1)

[46] *Speculative Remarks about Conservation and Unification* (Helv. Phys. Acta, 38, 1965, p. 98)

[47] *Les théorèmes de conservation et l'idée d'une théorie unitaire* (Atti del Convegno sulla Relatività generale, IV Centenario della nascita di Galileo Galilei, Firenze 1965)

[48] *Report on the Symposium on General Relativity and Gravitation held at the Berlin Academy* (Bull. GRG, No 11, 1966, p. 1)

[49] *Report on the New Delhi Seminar on Relativistic Cosmology and Gravitational Theories* (Bull. GRG, No 11, 1966, p. 1)

[50] *Le Colloque international du CNRS sur Fluide et Champ gravitationnel en Relativité Générale* (Bull. GRG, No 15, 1967, p. 1)

[51] (With H.-J. Treder) *Future of GRG (General Relativity and Gravitation)* (The Mathematical Student, XL, no 1, 1972, p. 49)

[52] *Time, the Concept of Gravitation and the Possibility of a Unified Field Theory* (Proc. International Symposium on *Relativity and Unified Field Theory*, Calcutta University Dec. 1975/Jan. 1976, p. 1)

[53] *Sur l'unification des théories physiques* (in *Gravitationstheorie und Theorie der Elementarteilchen,* Wiederabdruck ausgewählter Beiträge des Einstein-Symposiums 1965 in Berlin, im Auftrage der Einstein-Kommission beim Präsidium der Akademie der Wissenschaften von H.-J. Treder hg., Akademie-Verlag, Berlin 1979, p. 70-88)

[54] *Physical and Metaphysical Time* (Epistemologia, I, 1979, p. 337ss.)

[55] *Von der Erde zum Kosmos* (Die Sterne, 70, 1994, 5, pp. 14-26)

IX BOOK REVIEWS

1 FRIEDRICH DESSAUER, *Der Fall Galilei und wir* (Der Bund, Nr. 252 vom 1. Juni 1943, p. 3)

2 *Correspondance inédite entre Réaumur et Abraham Trembley* (Der Bund, Nr. 576 vom 9. Dez. 1943, p. 13)

3 RÉMY COLLIN, *Les deux savoirs* (Erasmus, 1, 1948, p. 336)

4 JEAN PELSENEER, *L'évolution de la notion de phénomène physique des primitifs à Bohr et Louis de Broglie* (Gesnerus, 5, 1949, p. 336)

5 P. GOMBAS, *Die statistische Theorie des Atoms und ihre Anwendungen* (Experientia, V, 1949, p. 488)

6 MARTIN JOHNSON, *Science and the Meaning of Truth* (Erasmus, 2, 1949, p. 336)

7 ANTONIO CARELLI, *Limiti e possibilità della Scienza* (Erasmus, 3, 1950, p. 417)

8 HANS THIRRING, *Die Idee der Relativitätstheorie* (Experientia, VI/5, 1950, p. 198)

9 HEIMO DOLCH, *Theologie und Physik* (Revue de Théologie et de Philosophie, t. III, 1953, p. 65)

10 ANDREAS SPEISER, *Elemente der Philosophie und der Mathematik* (Erasmus, 6, 1953, p. 138)

11 ERWIN NICKEL, *Das "physikalische Modell" und die "metaphysische Wirklichkeit". Versuch einer Metaphänomenologie* (Revue de Théologie et de Philosophie, t. III, 1953, p. 160)

12 GILBERT RYLE, *Dilemma* (Rev. de Théologie et de Philosophie, t. IV, 1954, p. 324)

13 FRANÇOIS MEYER, *Problématique de l'évolution* (Revue de Théologie et de Philosophie, t. V, 1955, p. 77)

14 PAULETTE FÉVRIER, *Déterminisme et indéterminisme* (Revue de Théologie et de Philosophie, t. VI, 1956, p. 77)

15 RAFFA & GONSETH, éd., *Philosophie néo-scolastique et philosophie ouverte* (Revue de Théologie et de Philosophie, t. VI, 1956, p. 154)

16 C. TRUESDELL, *The Kinematics of Vorticity* (ZAMP, VII, 1956, Fasc.6)

[17] PAUL SCHERRER. *Atomenergie* (Der Bund, Nr. 216 vom 10 Mai 1957)

[18] E. W. F. TOMLIN, *Living and Knowing* (Revue de Théologie et de Philosophie, t. VII, 1957, p. 157)

[19] EDMUND SINNOTT, *The Biology of the Spirit* (Revue de Théologie et de Philosophie, t. VII, 1957, p. 314)

[20] JEANNE VIAL, *Métaphysique* (Studia philosophica, XVII, 1957, p. 185)

[21] AMIR MEHDI BADI', *L'idée de la méthode des sciences*, 1: Introduction (Studia philosophica, XVII, 1957, p. 202)

[22] AGNES ARBER, *The Manifold and the One* (Revue de Théologie et de Philosophie, t. VIII, 1958, p. 239)

[23] MAX PLANCK, *Physikalische Abhandlungen und Vorträge* (Chimia, 12, 1958, p. 366)

[24] *Logica Studia* PAUL BERNAYS *dedicata* (Erasmus, 13, 1960, p. 68)

[25] A. NEUHAEUSLER, *Zeit und Sein* (Erasmus, 14, 1952, p. 643)

[26] AMIR MEHDI BADI', *L'illusion de l'extensibilité infinie de la vérité. II: Vers une connaissance objective* (Studia philosophica, 22, 1963, p. 259)

[27] R. M. MCINERY, *The logic of Analogy* (Erasmus, 15, 1963, p. 344)

[28] VIKTOR KRAFT, *Erkenntnislehre* (Erasmus, 15, 1963, p. 392)

[29] M. SCHRAMM, *Ibn al-Haythams Weg zur Physik* (Erasmus, 16, 1964, p. 259)

[30] E. ANRICH, *Moderne Physik und Tiefenpsychologie* (Erasmus, 17, 1965, p. 193)

[31] E. LASZLO, *Essential Society* (Erasmus, 17, 1965, p. 198)

[32] K. M. MEYER-ABICH, *Korrespondenz, Individualität und Komplementarität* (Erasmus, 18, 1966, p. 132)

[33] H. STACHOWIAK, *Denken und Erkennen* (Erasmus, 19, 1967, p. 146)

[34] DESCARTES, *Œuvres philosophiques*, t. I, textes établis par E. ALQUIÉ (Erasmus, 19, 1967, p. 580)

[35] K. H. BURMEISTER, *Rheticus*, Bd. I (Erasmus, 19, 1967, p. 713)

[36] DEMETRIO DA CREMA, *Mondo esterno nel Zamboni* (Erasmus, 19, 1967, p. 715)

[37] W. BOEHM, *Johannes Philoponos, Grammatikos von Alexandrien* (Erasmus, 20, 1968, p. 197)

[38] ANTON FISCHER, *Die philosophischen Grundlagen der wissenschaftlichen Erkenntnis* (Erasmus, 21, 1969, p. 202)

[39] K. H. BURMEISTER, *Rhetikus*, Bd. II & III (Erasmus, 21, 1969, p. 257)

[40] SUZANNE SCHMIDA, *Perspektiven des Seins* (Erasmus, 21, 1969, p. 389)

[41] R. H. KARGON, *Atomism in England* (Erasmus, 21, 1969, p. 458)

[42] Akten des Internationalen Leibniz-Kongresses, Vol. I-II (Erasmus, 22, 1970, p. 1)

[43] C. B. SCHMITT, *Gianfrancesco Pico della Mirandola* (Erasmus, 22, 1970, p. 260)

[44] D. B. RICHARDSON, *Berdyaev's Philosophy of History* (Erasmus, 22, 1970, p. 327)

[45] VIKTOR KRAFT, *Die Grundlagen der Erkennntnis und der Moral* (Erasmus, 22,1970, p. 385)

[46] GERARD RADNITZKY, *Contemporary Schools of Metascience* (Journal of the British Society for Phenomenology, 1, no 3, 1970, p. 97)

[47] M. DESCHOUX, *Philosophie du savoir scientifique* (Erasmus, 23, 1971, p. 129)

[48] H. VETTER, *Wahrscheinlichkeit und logischer Spielraum* (Erasmus, 23, 1971, p. 138)

[49] R. BOEHM, *Vom Gesichtspunkt der Phänomenologie* (Journal of the British Soc. for Phenomenology, 2, 1971, p. 92)

[50] J. P. LEYVRAZ, *Phénoménologie de l'expérience* (Erasmus, 23, 1971, p. 586)

[51] *Wittgenstein et le problème d'une philosophie de la science.* Publ. du CRNS (Erasmus, 23, 1971, p. 589)

[52] M. L. VON FRANZ, *Zahl und Zeit* (Erasmus, 23, 1971, p. 970)

[53] G. G. GRANGER, *Essai d'une philosophie du style* (Journ. Brit. Soc. of Phenomenology, 2, 1971, p. 91)

[54] C. DE WAARD, *Correspondance de Mersenne* (Erasmus, 24, 1972, p. 75)

[55] G. CRAPULLI, *Mathesis universalis* (Erasmus, 24, 1972, p. 129)

[56] MICHAEL WOLF, *Fallgesetz und Massebegriff. Zwei wissenschaftshistorische Untersuchungen zur Kosmologie des J. Philoponus* (Revue de Théologie et de Philosophie, t. 22, 1972, p. 138)

[57] P., ENGELMANN, *Ludwig Wittgenstein. Briefe und Begegnungen* (Erasmus, 24, 1972, p. 449)

[58] R. G. COLODNY, *Scientific Theories, Essays etc.* (Erasmus, 24. 1972, p. 519)

[59] G. SCHISCHKOFF, *Peter Beron* (Erasmus, 24, 1972, p. 647)

[60] N. RESCHER & A. URQUHART, *Temporal Logic* (Erasmus, 25, 1975, 1973, p. 6)

[61] VIKTOR KRAFT, *Mathematik, Logik und Erfahrung* (Erasmus, 25. 1973, p. 711)

[62] *R. Carnap in Memory* (Erasmus, 25, 1973, p. 257)

[63] K. ASCHENBRENNER, *Concepts of Value* (Erasmus, 25, 1973, p. 581)

[64] M. BLONDEL, *Lien substantiel et substance composée* (Erasmus, 25, 1973, p. 711)

[65] IVOR LECLERC, *Physical Existence* (Erasmus, 25, 1973, p. 716)

[66] Akten des Intern. Leibniz-Kongresses, Bd. III, IV, V (Erasmus, 25, 1973, p. 769)

[67] H. SCHIEDERMAIR, *Macht und Recht bei Leibniz* (Erasmus, 26, 1974, p. 69)

[68] J. HOROWITZ, *Law and Logic* (Erasmus. 26, 1974, p. 135)

[69] Studia Leibniziana, Bd. VI & VII (Erasmus, 26, 1974, p. 263)

[70] W. K. ESSLER, *Wissenschaftstheorie* (Erasmus, 26, 1974, p. 389)

[71] J. R. LUKAS, *Treatise on Time and Space* (Erasmus, 26, 1974, p. 460)

[72] W. K. ESSLER, *Induktive Logik* (Erasmus. 26. 1974. p. 716)

[73] MAX FLÜCKIGER, *Einstein in Bern* (Der Bund: *Der kleine Bund,* Nr. 263, vom 10. Nov. 1974)

[74] F. KRAFFT et al., Internationales KEPLER-Symposium, (Erasmus, 278, 1975, p. 331)

75 F. SUPPE, Editor, *Structure of Scientific Theories* (Erasmus, 27, 1975, p. 459)

76 ROMAN INGARDEN, *Kausale Struktur der realen Welt* (Erasmus, 27, 1975, p. 706)

77 D. I. BLOKHINTSEV, *Space and Time in the Microworld* (Erasmus, 28, 1976, p. 129)

78 M. STRAUSS, *Modern Physics and its Philosophy* (Erasmus, 28, 1976, p. 134)

79 G. BÖHME, *Zeit und Zahl* (Erasmus, 28, 1976, p. 198)

80 V. BERNING, *Wagnis der Treue* (Erasmus, 28, 1976, p. 262)

81 J. P. SCHOBINGER, *Pascals "De l'esprit géométrique" und "De l'art de persuader"* (Erasmus, 28, 1976, p. 524)

82 W. STEGMÜLLER, *Theorie und Erfahrung* (Erasmus, 29, 1977, p. 138)

83 B. BOLZANO, *Kleine Wissenschaftslehre* (Erasmus, 30, 1978, p. 65)

X PREFACES TO BOOKS BY OTHER AUTHORS

1 Vorwort zu VIKTOR GORGÉ, *Philosophie und Physik* (Verlag Duncker & Humblot, Berlin 1959)

2 Foreword to J. P. MARCHAND, *Distributions, An Outline* (North Holland Publ. Co., Amsterdam 1962)

3 Vorwort zu GOTTFRIED T. RÜTTIMANN, *Logikkalküle der Quantenphysik* (Verlag Duncker & Humblot, Berlin 1977)

XI EULOGIES and OBITUARIES

1 *Arthur Schidlof*† (Actes, Soc. Helv. des Sci. Naturelles, 1935, p. 484)

2 *Zum Tode Jean Perrins* (Der Bund, 1944)

3 *Paul Grunert*† (Actes, Soc. Helv. des Sci. Naturelles, 1958, p. 363)

4 *Professor Wolfgang Pauli zum Andenken* (Der Bund, Nr. 567 vom 23. Dez. 1958, p.1-2)

Idem (Der Aufbau, vol. XXV, no 10, New York, March 6, 1959)

5 *Professor Robert Schafrot zum Andenken* (Der Bund, Nr. 232 vom 4. Juni 1959, p. 2)

6 *Niels Bohr*†(Industries atomiques, 6, Nos. 11-12, 1962, p. 100)

7 *Eloge de Hans Barth* (Studia philosophica, XXV, 1965, p. 12)

8 *In memoriam Christian Møller* (GRG-Journal, Vol. 12, No 1, 1980, p. 91-92)

9 *Louis de Broglie, prix Nobel de physique* (in *Louis de Broglie que nous avons connu.* Bibliothèque des Annales de la Fondation Louis de Broglie. Conservatoire National des Arts et Métiers, Paris 1988, pp. 149-150)

10 *Alexander von Muralt, 1903-1990* (Archives des Sciences, Genève, Vol. 45, Fasc. 1, 1992, pp. 103-104)

XII POETRY (under a pseudonym)

1 ARN DE MONTERNE, *De l'amour et du temps, Bribes de métaphysique* (© at the Author's, numbered edition, Berne 1968, pp. 1-94)

XIII OTHER LITERARY ACTIVITY

1 *Blaise Pascal, Erfinder der Rechenmaschine* (Der Bund, Januar 1943)

2 *"Die Physik des Nichts" nach einem Vortrag von Karl Popper* (Der Bund, Nr. 604 vom 28. Dez- 1946, p. 2)

3 *Courte autobiographie* (Acad. friburgensis, VIII/1, 1949/50, p. 22)

4 *Der dänische Gelehrte Niels Bohr und die Atomphysik* (Kontakt mit Dänemark, Nr. 3, 1950, p. 8)

5 *La Conférence internationale des universités à Nice, 4-10 décembre 1950* (Revue universitaire suisse, 24, 1951, p. 7)

6 Thèses présentées sur *L'enseignement scientifique* au Deuxième Cours d'information sur les problèmes de l'UNESCO, 6-13 octobre 1951 (Textes français et allemand, polygr. Distribution limitée)

7 *A propos d'un film sur la physique atomique* (C.R. des séances de la Section des Sci. nat. et math., Bull. Institut National Genevois, VII, 1953, p. 13)

8 *Europäische Zusammenarbeit auf dem Gebiete der kernphysikalischen Forschung* (Schweizerische Hochschulzeitung, 25, 1952, Nr. 2, p. 135)

9 *A propos du Laboratoire nucléaire européen* (Journal de Genève du 17 octobre 1952, p. 1)

[10] *Das europäische Kernforschungslaboratorium* (Der Bund, Nr. 27 vom 18. Januar 1953)

[11] *Impressions bernoises* (Journal de Genève, Suppl. du No 21, p. 1, du 26 janvier 1954)

[12] *Das Atom für den Frieden* (Der Bund vom 11. August 1955, p. 6)

[13] *Le Collège et le Gothique américain* (Journal de Genève du 16 novembre 1955, p. 1)

[14] *"Smog", Palomar et Santa Fé* (Journal de Genève du 8 février 1956, p. 3)

[15] *La plus belle ville du monde* (Journal de Genève du 26 mars 1956, p. 3)

[16] *Eine amerikanische Universität* (Der Bund vom 10. Januar 1956, p. 2)

[17] *In Warschau wurde der Marxismus erörtert* (Der Bund Nr. 395 vom 26. August 1957)

[18] *Niels Bohr und der Frieden* (Der Bund Nr. 501 vom 27. Oktober 1957, p. 2)

[19] *Les sciences en Suisse romande* (Cahiers de la Fondation Plisnier, 1958, p. 130)

[20] *Der Begründer der neuen Physik. Zum 100. Geburtstag von Max Planck am 23. April 1958* (Der Bund, Nr. 185 vom 23. April 1958, p. 1-2)

[21] *Journal d'Asie* (Polygr, Chez l'Auteur, Berne 1959)

[22] *En hilsen fra Udlandet til Kontorchef Bj. Gamborg's 50 års jubilæum i Akademisk Orkester København* (Optakten, Nr. 43, Jan. 1962, p. 7)

[23] *Discours d'ouverture* du Congrès de l'Association Internationale d'Hydrologie scientifique, Année Géophysique des Nations Unies, par le Recteur de l'Université de Berne (Publ., Intern. Assoc. of Scientific Hydrology, XII Year, 1967, p. 121)

[24] *"Herr Mercier macht sich Gedanken". Probleme der zeitgenössischen Universität, vom Rektor der Universität dargelegt* (Berner Student, 36. Jg., Nrn. 8/9/10, 1968)

[25] *Les arts et les sciences, et le statut des Ecoles de musique comparé à celui des Facultés universitaires* (Actes du 6e Congrès triennal de l'Association Européenne des Académies, Conservatoires et Hautes Ecoles de Musique, Berne 1969, p. 21)

[26] *Institut für exakte Wissenschaften: Mehr Raum* (Der Bund, Nr. 127 vom 4. Juni 1979, p. 15)

[27] (Under a pseudonym) ARN DE MONTERNE, *Jolly-sur-Rabretière* - (Chez l'Auteur, Berne et la Rabretière du Bernard 1972. Seconde impression sans changement 1975, p. 1-45)

[28] Entrevista do Prof. ANDRÉ MERCIER (Soc. Brasil. de Filos. Católicos, Boletim No 4. Ano 1973)

[29] *"Science has become aggressive"* (in Russian) (Interview in the Literary Gazette of Moscow, No 50 of Dec. 11, 1974, p. 13. With a *Replique* by B. LEKTORSKY and a *Conclusion* by F. KONSTANTINOV)

[30] Discussion II on: J.-CL. PIGUET, *Mathématiques et musique* (Jahresbericht der Schweizerischen Geisteswissenschaftlichen Gesellsch. 1974, p. 244)

[31] Discours d'ouverture: *Marksistowska Krytyka Fenomenologii a Filoozofia* ROMANE INGARDEN (Z Historii Filosofii, Studia Filosofoczne, Nr. 9(118), 1975, p. 8)

[32] *Anabase*, à SAVA GANOVSKI, Président de la FISP, pour ses 80 ans (Suppl. au Bull. de la FISP, Fédération Internationale des Sociétés de Philosophie, Berne 1977)

[33] *De l'art de l'édition - Propos d'un amateur* (in *Einheit in der Vielheit*, Festschrift für PETER LANG zum 60. Geburtstag, Verlag Peter Lang, Bern-Frkft/M.-New York-Paris 1988, pp. 312-325)

[34] *Le violon: Œuvre d'art, ou d'artisanat? Contribution à la théorie esthétique du "viol des symétries"*. Essai offert à EDITH DITTRICH, Maître Luthier, à l'occasion des Trente années passées à diriger l'atelier de lutherie créé par son grand-père soixante ans plus tôt (© Chez l'Auteur. Université de Berne 1990, pp. 1-31)

[35] *Wozu Jubiläumsfeiern?* (in *The Earth and the Universe*, A Festschrift in honour of H. J. TREDER, ed. by W. SCHRÖDER, Interdiv. Comm. on History of the Internat. Assoc. of Geomagnetism and Aeronomy (IAGA), Bremen-Rönnebeck (Germany), 1993, pp. 293-296)

- o -